AF517863

Integrated Aquaculture

Dr. B. Ahilan is working as Professor in the Department of Aquaculture, Fisheries College and Research Institute, Thoothukudi. He did his Ph.D., in Aquaculture and has put up a total service on 21 years in the university. He has completed a Part-II research project on "Poultry cum fish culture". He has published 4 text books, 35 research papers and 86 popular articles related to aquaculture. He has organized and co-ordinated several training programmes in fish culture for the benefit of fish farmers, rural folks and entrepreneurs.

Dr. K. Ravaneswaran is working as Associate Professor of fisheries in Fisheries Research and Extension Centre at Tamil Nadu Veterinary and Animal Sciences University. He has put up a total service of 20 years in the university. He has established live feed culture unit and ornamental fish breeding unit. He has published 10 research papers, 70 popular articles, 1 book and 11 training manuals.

Dr. P. Kumaravel, Ph.D., Associate Professor is presently working as Programme Coordinator cum Head, Krishi Vigyan Kendra, Kattupakkam under Tamil Nadu Veterinary and Animal Sciences University. He did his Masters and Doctorate in the discipline of Veterinary and Animal Husbandry Extension and has put up a total service of 13 years in the university. He has a total of 165 publications to his credit which includes 19 research papers, 68 popular articles and papers, 29 books/manuals, and 48 pamphlets. He has developed three interactive softwares on animal welfare, package of animal husbandry and poultry practices and scientific pig farming.

Integrated Aquaculture

Dr. B. Ahilan, *M.F.Sc. Ph.D.,*
Professor
Department of Aquaculture
Fisheries College and Research Institute
Tuticorin – 628 008
Dr. K. Ravaneshwaran, *M.F.Sc. Ph.D.,*
Associate Professor
Fisheries Unit
Mathavaram Milk Colony
Chennai – 600 051
Dr. P. Kumaravel, *M.V.Sc. Ph.D.,*
Associate Professor
Livestock Research Station
Kattupakkam

2026
Daya Publishing House®
A Division of
Astral International Pvt. Ltd.
New Delhi – 110 002

© 2026 AUTHORS
First Published, 2011
Reprinted, 2021
ISBN: 978-81-7035-679-0

Disclaimer:

Every possible effort has been made to ensure that the information contained in this book is accurate at the time of going to press, and the publisher and author cannot accept responsibility for any errors or omissions, however caused. No responsibility for loss or damage occasioned to any person acting, or refraining from action, as a result of the material in this publication can be accepted by the editor, the publisher or the author. The Publisher is not associated with any product or vendor mentioned in the book. The contents of this work are intended to further general scientific research, understanding and discussion only. Readers should consult with a specialist where appropriate.

Every effort has been made to trace the owners of copyright material used in this book, if any. The author and the publisher will be grateful for any omission brought to their notice for acknowledgement in the future editions of the book.

All Rights reserved under International Copyright Conventions. No part of this publication may be reproduced, stored in a retrieval system, or transmitted in any form or by any means, electronic, mechanical, photocopying, recording or otherwise without the prior written consent of the publisher and the copyright owner.

Published by : **Daya Publishing House**®
A Division of
Astral International Pvt. Ltd.
– ISO 9001:2015 Certified Company –
4736/23, Ansari Road, Darya Ganj
New Delhi-110 002
Ph. 011-43549197, 23278134
E-mail: info@astralint.com
Website: www.astralint.com

Digitally Printed at : **Replika Press Pvt. Ltd.**

TAMIL NADU VETERINARY AND ANIMAL SCIENCES UNIVERSITY

Dr. N. Daniel Joy Chandran, ***Ph.D.,***
Registrar

Madhavaram Milk Colony
Chennai – 600 051

Foreword

Integrated fish farming with livestock and poultry is gaining momentum all over the world owing to the recycling of wastes generated in this system for fish production besides abating pollution. As there is no expenditure towards feed and fertilizer, the cost of fish production in integrated farming systems is comparatively low. Further such eco-friendly culture systems bring additional income and social benefits to the rural community. The different technologies relating to integrated systems need to be popularized in India so as to make use of the wastes of livestock and poultry for beneficial purposes. A book on Integrated Aquaculture is the need of the hour to help the students of Fisheries Science in State Agricultural Universities and in conventional universities where fisheries is taught as a subject. Keeping these in view, this book is brought out in a readable style and I strongly hope that this publication would be of immense use for different constituents of fisheries sector.

I congratulate the authors of this publication for their sincere efforts.

N. Daniel Joy Chandran

Preface

Fisheries play a vital role in feeding the world's population, contributing significantly to the dietary protein intake of hundreds of millions of people. The rationale of aquaculture is not limited merely to socio-economic and marketing advantages. Irrespective of the economic or other benefits of large-scale aquaculture operations, greater emphasis is laid to small scale farming in developing countries. Integrated aquaculture can compliment and improve the overall efficiency of many types of farm. In order to balance the risks in farming and to increase the production and income without imposing adverse effects on their environment, farmers will have to adopt integrated farming practices.

Keeping the importance of integrated farming practices in the coming years, this book is prepared. This book emphasizes different integrated fish farming practices, nutrient dynamics, chemical composition of animal wastes and economics of different integrated fish farming systems.

The intended readership is primarily undergraduate and postgraduate students of aquaculture. It is also hoped that aquaculturists will also found this book useful.

We are highly indebted to Dr. N. Daniel Joy Chandran, Registrar, Tamil Nadu Veterinary and Animal Sciences University, Chennai for his Foreword. We are thankful to Dr. M. Venkatasamy, Ph.D., Dean i/c, Fisheries College and Research Institute for his kind help in the preparation of this publication. We extend our thanks to Mr. T. Chermaraj for his secretarial assistance.

Authors

Contents

Chapter 1
Introduction

The cost of land and construction are the major capital investments in a fish farm. These costs are increasing throughout the world and in addition other costs for fish and feed are also rising alarmingly. It is therefore important that fish (and other animal protein) production per unit area needs to be increased to help offset those increasing costs and also to help mitigate the world's "protein – hunger" prevalent in developing countries. One best way to achieve this objective is a combination of livestock/ poultry enterprises with fish farming.

Integrated farming may be defined as a sequential linkage between two or more farming activities. When the fish becomes a major commodity of this system it is known as integrated fish farming. The integration of aquaculture with livestock and crop farming offers great efficiency in resource utilization, reduces risk by diversifying crop, and provides additional food and income. This system involves recycling of waste or by-products of one farming system which in turn serves as an input for another system and efficient utilization of available farming space for maximum production. Integrated fish farming is well developed in China, Hungary, Germany and Malaysia, and is accepted as a sustainable form of aquaculture and major contributor of farmed fish. Freshwater aquaculture in India is organic-based and derives inputs from agriculture and animal husbandry. India possesses the largest bovine population of over 307 million cattle herds, along with 181 million sheep and goats, 16 million pigs, and over 150 million poultry and other livestock. India being an agrarian economy, produces large quantities of plant and animal residues, to the tune of over 322 and 1,000 million metric tones, respectively on an annual basis. In addition activities like mushroom cultivation, rabbit, sericulture and apiculture, provides huge quantities of organic material for aquaculture. Agro-based industries like distilleries and food-processing plants also produce effluents which could be recycled for aquaculture in addition to domestic sewage to the extent of over 4,000 million litres on daily basis.

Aquaculture and agriculture compliments and supplements each other by balancing the economy of natural resources. Pond embankments could be used for growing napier grass for grass carp. Vegetables such as tomatoes grow well on bundhs, which are fertilized with pond silt rich in plant nutrients. Pond silt from fertile ponds, in many ways, are equal in value to good quality compost and can be used us a fertilizer in gardens and field. The pond embankments could also be planted with fruit trees, which provide a regular supplemental income. By – products of the agriculture industry such as wheat and rice-bran and oil cakes could be profitably utilized in aquaculture in rural areas where they are cheap and plenty.

The two vocations namely aquaculture and animal husbandry can also aid and complement each other. The use of domestic animal wastes for fertilizing fishponds leads to higher fish yields as it provides almost all the raw materials required for the metabolic cycle in the pond. Cattle, sheep, goat and pigs could be raised on pond embankments or in their vicinity. Pigs have a special significance and are considered as "costless fertilizer factories moving on hooves" and pig manure constitutes the main source of homemade fertilizer. Pond embankments, which normally have a perennial green vegetative cover of grasses, provide ideal pastures for animal grazing. Poultry farming and aquaculture could be effectively combined, since poultry litter serves as a fertilizer for ponds while the offal's of fish are effectively utilized as feed for poultry. Duck raising in ponds and on pond embankments has a special advantage, in that they feed on snails which serve as vectors of certain infectious diseases and thus reduce their incidence considerably.

The linkages within integrated fish farms are numerous and highly intricate. Livestock manures usually derived from animal rearing within the farm and night soil of humans are used directly as pond fertilizers for the microbial food web. Residues from sericulture, including silkworm droppings, cocoon processing wastes water and pupae serve as pond inputs. Agricultural by – products like dregs from wine processing, wheat and rice chaff, cotton or rapeseed meal (after oil extraction) and other locally available by-products are also utilized into the ponds

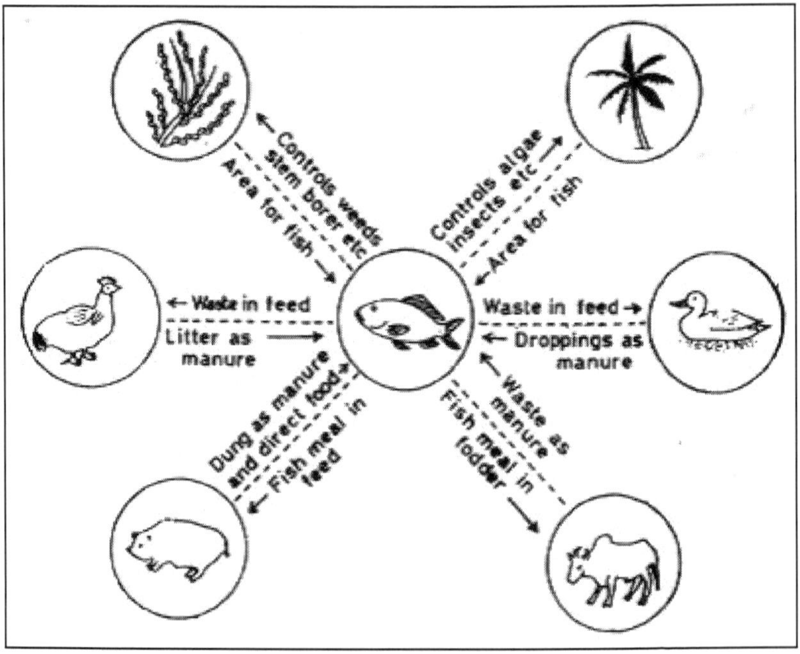

Figure 1: Integrated Agriculture, Aquaculture and Animal Husbandry

as fish feed. In addition floating aquatic plants; plants grown on ponds and canals can be utilized as green fodder for fish and livestock. The pond bottom humus is also employed as fertilizer for crops grown adjacent to ponds or directly on their dykes, including fish feeds (green fodders) or human foods (Vegetables and fruits). The farms are designed to optimize the use of wastes from each process directly and to use the available space most efficiently. The significant advantage of utilization of such a wide variety of wastes helps to control pollution through maximal recycling of many waste materials that would otherwise need to be discarded.

One of the distinct feature of integrating agriculture, aquaculture and animal husbandry is various diseases like malaria, filariasis, dengue, guinea worm and encephalitis could be effectively controlled. It can also directly aid in the production of both animal and plant food and play a significant role in the community health especially in rural areas. On integrating aquaculture it offers a great scope of balancing the rural economy in the country by utilizing these waste products and in turn providing quality protein.

Integrated fish farming is an ancient practice in China and the immigrant Chinese have introduced it in several Southeast Asian countries. It can be carried out without a great deal of additional expense and with minimum adverse effect on crop and animal farming. It is a means of diversifying farm outputs and producing food for the rural folk. The labour required can be easily shared between family members and the farm wastes utilized for fertilizing and feeding the fish.

It is a means of land reclamation in certain areas and the relatively wide embankments built in such pond farms serve to increase the total land area available for crop and animal farming. Besides these such integrated farming can play in increasing the employment opportunities and it has received considerable attention in recent years. Besides many developing countries of Asia, some in Africa and South America have introduced this system on a pilot or larger scale. The basic principles involved in integrated farming are the utilization of the synergetic effects of inter-related farm activities and the conservation, including the full utilization, of farm wastes. It is based on the concept that 'there is no waste' and 'waste is only a misplaced resource which can become a valuable material for another product' (FAO, 1977).

Fisheries play a vital role in feeding the world's population, contributing significantly to the dietary protein intake of hundreds of millions of people. Fish supplies about 30 per cent of the total animal protein in the diet of Asian population, 20 per cent in Africa and 10 per cent in Latin America" (FAO, 1996). Worldwide aqua food consumption increases with income (Born *et al.*, 1994). It is also reported that in a number of countries with the lowest average incomes, the ration of aqua food consumption to meat consumption rises dramatically, and in many cases more fish is consumed than meat.

A fishpond is a semi-closed biological unit from the biological production point of view. The pond provides not only space for fish but also offers oxygen for respiration, natural foods and other requirements. Monoculture of a fish species can never utilize all the pond's biological resources. Conversely, the adoption of polyculture (the combined culture of two or more fish species) enables a more effective utilization of the resources of various food niches in the pond.

Today living aquatic resources are intensely harvested around the world as a source of protein in the family diet, for commercial gain or for recreation. Aquatic farming has a special significance in fish marketing strategies. Production can be organized according to market demand with respect to quantity, preferred size, colour, preservation and processing etc. In many markets, there is a special demand for fresh or chilled fish and it may not be easy for the fishing industry to adequately satisfy such a demand. Harvesting from farms can be regulated to meet this demand and the product may be made available during off seasons in order to maintain regular supplies. Also, the species can be grown to the size most preferred by consumers, when size restrictions have to be observed in capture fisheries.

The rationale of aquaculture is not limited merely to socio-economic and marketing advantages. Irrespective of the economic or other benefits of large-scale aquaculture operations, greater emphasis is laid on small scale farming in developing countries. The three promising aquaculture systems, which hold promise for meeting the global food requirements are culture of fishes, culture of filter – feeding molluscs and integrated fish farming.

Fish production in organic manured ponds depends on the availability of nitrogen and phosphorus as well as on the amount of organic carbon introduced and produced. With the help of the bound biological energy, nutrients are converted to fish biomass by various organisms of the aquatic food web. In the intensely manured fishponds, the single management input to supply nutrients to the food web is the manure. Organic fertilization feeds both the autotrophic algal and heterotrophic bacterial detrital food chains, which in turn supply food to fish for growth directly or indirectly.

Supplementary feeds and fertilizers play a vital role in increasing fish production in aquaculture operations, but as supplementary feeds are scarce and costly, the application of either organic or inorganic fertilizers is a low cost alternative for fish culture. Since the inorganic fertilizers, which are used extensively in agriculture, have resulted in an increasing gap between demand and supply, excreta could be used to fertilize fishponds.

Continued growth in aquaculture is only possible if the relative availability of its primary inputs namely land, water and feeds can be increased. In many developing areas, suitable land and water is already used for agriculture and therefore, further aquaculture development will depend on profitability of aquatic products or as an integrated enterprise with the existing one.

Sustainability Factors

(*i*) Water

A key factor in the sustainability of Inland aquaculture is the water availability since it is already in high demand for agriculture. Carp production in Asia, accounting for almost half of the global production in aquaculture (excluding plants), relies on long standing extensive and semi-intensive polyculture systems, which are extremely efficient in the use of water (FAO, 1995). Further, expansion of this form of aquaculture is possible without competition for water resources, by integrating fish farming into waters that lend themselves to multiple uses.

(*ii*) Land

In general, the conditions which favour agriculture, namely relatively flat topography, abundant water and long growing season, also favour aquaculture. Therefore, except in low-lying areas or those with saline soils, pond culture will have to compete with agriculture for land use. Its ability to compete will depend on its effectiveness for reducing small farm risk, producing income, and maximizing socio-economic benefits. Expansion of aquaculture is possible on land that is already given to agriculture by integrating the two production systems. The loss of agriculture production is made up by a number of potential benefits such as the production of fish ponds, the opportunity for assimilation of agricultural wastes, the availability of pond sediment for soil improvement, modification of the water table, increased diversity of crops and the social benefits of rural employment opportunities.

Integration at the policy level must go hand in hand with the farm level integration of fish farming. Where fish farming and agriculture intersect, there is a need for policy to regulate land and water use and to assess environmental impacts, with the objective of maximizing the sustainability and carrying capacity of farming systems. Opportunities can also be found for integrating aquaculture with agricultural infrastructure such as facilities for feed production, fertilizers supplies and extension services. When there is strong competition for the use of land, water or other resources, there will be a need to integrate aquaculture into local planning activities, particularly with respect to site selection, water supplies and environmental impact assessment.

It is apparent that semi-intensive integrated farming systems can be far less polluting than more intensive systems. It is always better to go for semi-intensive integrated farming systems for several reasons: their low impacts on the environment, their low fish production costs and their protein production efficiency which is primarily due to the efficiency with which they convert nutrients and energy. Compared to semi-integrated farming practices intensive fishponds were more polluting. Multi occupation integrated fish farming network not only increases the sources of feed and fertilizer available while lowering their costs, but also increases the variety of produce available and enhances production value.

Integrated Agriculture-Aquaculture (IAA) has been widely and vigorously promoted as a means for addressing issues of rural hunger and poverty. The principal objective of promoting IAA is to enhance the production of fish and other crops for human consumption through the synergistic interaction of pond, field and livestock, upgrading the household nutritional status of farm families and/or increasing cash or in-kind income. This system of farming utilizes wastes from different components, livestock, poultry and agriculture by-products for fish production. Organic wastes to the tune of 40-50 kg are converted into 1 kg of fish, while pond silt is utilized as fertilizer for fodder-crop; which in turn is used to raise livestock and poultry or as fish feed, thus achieving complete recycling of waste. The scope of integrated farming is considerably wide. Ducks and geese are raised in pond, and pond-dykes are used for horticultural and agricultural crop production and animal raising. The system provides meat, milk, eggs, fruits, vegetables, mushroom, fodder and grains, in addition to fish. It utilizes water body, water surface, land and pond silt for increased food production economically for human consumption. It holds a great promise and potential for augmenting production, betterment of rural economy, employment generation and also improving socio-economic status of weaker sections of rural community.

Chapter 2

Current Status and Prospects of Integrated Fish Farming in India

Fish farmers in India are attempting to optimize their use of farm resources in order to maximize their returns. The cost of fish feed and pond fertilizers in aquaculture accounts for approximately 60 per cent of total farm expenditure and these costs can be sharply reduced by integrating this system with other farming systems. Fish culture in India has undergone gradual change during the past few decades and there has recently been some movement to integrate livestock and crop farming. Nonetheless, the sustained research efforts of scientists in India to develop production systems that would optimize farmers' returns through the judicious use of farm wastes have resulted in integrated farming system in which livestock such as ducks, pigs and poultry are raised along with fish.

The benefits of integrated fish farming extend beyond improved fish production and include an increase in the production of livestock and eggs. Fodder and other horticultural crops grown on terraced embankments of fish ponds, for example, are used to feed livestock and fish and are also consumed by farm families, while the increase in costs incurred to raise additional animals is offset by the sale of meat and eggs. Apart from providing fish, cereal, fruits, milk, eggs and meat, a mixed farming system also provides a new way of life for the farmers.

Important Integrated Fish Farming Systems

Aquaculture–Agriculture integration

(*a*) Rice-fish integrated farming

(*b*) Rice-azolla-fish integration

(*c*) Fodder-cattle-fish integration

(*d*) Horticultural crop-fish integration

(*e*) Mulberry-silkworm-fish integration.

Aquaculture–Livestock integration

(*a*) Fish-cattle integrated farming

(*b*) Fish-pig integrated farming

(*c*) Fish-goat/sheep integrated farming

(*d*) Fish-poultry integrated farming

(*e*) Fish-duck integrated farming

Resources and Potential

Freshwater aquaculture resources of the country are vast, (2.15 million hectares of ponds and tanks and 1.3

million hectares of beels and derelict water, in addition to 1.2 million km of canals and three million hectares of reservoirs) that can be put into different fish culture practices. This sector with its present contribution of 1.38 million tonnes of fish/shell fish worth over Rs.4000 crores has a potential of producing over 4.5 million tonnes annually if the available technologies are transferred and fully adopted. In terms of cultivable species of fish/shell fish, the components are diverse to suit the ecological conditions of different water bodies to meet the regional preferences.

Status of Paddy-Fish Integrated Farming in India

The paddy-fish integrated farming is being adopted in areas where paddy fields retain water for three to eight months in a year and in fields which remain flooded even after the paddy is harvested and this could also serve as an off-season occupation for farmers. A traditional paddy cum prawn culture system is popular in the paddy fields of adjoining tidal backwaters and estuaries of rivers. During the monsoon months of June to September, when the water in the field is almost fresh, paddy is cultivated and during the rest of the year, the fields are utilized for prawn culture operations. There is about 26,000 hectares of total brackish water of which only 5000 hectares are currently utilized for growing paddy during the monsoon season and prawn during the rest of the year. Thus from 5000 hectares of paddy fields in central Kerala, approximately 6000 tonnes of prawn catch is estimated in addition to paddy crop. Moreover, in brackish water areas of west Bengal and Kerala where paddy fields are traditionally used to culture prawns, production of 50 kg/ha of *Penaeus monodon*, 250kg/ha of mullets and 3000 kg of tilapia have been achieved in addition to 2.4 ton/ha of paddy.

Many experiments have been conducted in the North-eastern and Southern States of India which revealed that the culture of fish in paddy fields was generally useful to paddy due to the better aeration of water and a greater tillering effect caused by the movement of fish in the fields. Fish excreta improved the soil's fertility and the introduction of herbivorous fish helped to control weeds and reduced weeding labour and costs. On the other hand, paddy fields provided a much poorer environment for the fish than did ponds, because fields were extremely shallow and subject to wide temperature and oxygen fluctuations, adverse pH, low phosphate and poor plankton production.

Status of Livestock–Fish Integrated Farming in India

A series of experimental trials were made under the Operational Research Project (ORP) in India (West Bengal), integrating fish culture and livestock farming practices. This farming system, which has also been tested under field conditions, can be modified for adoption in appropriate areas. Experiments and field trials involving pig, duck and poultry integrated with fish has revealed encouraging results. For example, it was observed, that the excreta of 35–40 pigs, 200–300 ducks and 250–300 layer broiler birds produced 6–7 tonnes, 3–4 tonnes, and 4 tonnes of fish per year respectively when recycled in one ha of water area under the polyculture of Indian and exotic fish. Among the carps, *Hypophthalmichthys molitrix* (silver carp) recorded the best growth, followed by *Ctenophayngodon idella* (grass carp), *Labeo rohita* (rohu), *Cirrhinus mrigala* (mrigal) and *Cyprinus carpio* (common carp). The per cent net returns were 75.0, 69.3 and 50.3 respectively for fish cum pig, fish cum duck and fish cum poultry farming. The costs of fish productions in these trials were much lower than that of conventional fish farming.

Prospects of Integrated Fish Farming with Reference to Indian Context

Fish-Paddy Integrated Farming

The reasons for the failure of fish-paddy integrated farming are attributed to poor or low water retention in paddy fields; the use of insecticides on dwarf high yielding varieties, harvested fish of relatively low weight and low market value. The adoption of high technology agricultural practices in this instance has served to reduce the viability of fish-paddy integrated farming. The future prospects for fish-paddy integrated farming will improve only if the agriculturists come to terms with the pisciculturists and use those pesticides which are compatible with fish to combat insect infestations. The International Rice Research Institute (IRRI), Philippines is developing certain strains of rice which are highly disease resistant. In addition biological means of controlling pests and other predators are also being developed as part of a move toward integrated pest management. The technology of fish-paddy integrated farming is not widely understood and much more research is needed to determine the optimum choice of fish species, size, stocking rate, mix of species and feeding practices.

Fish-Dairy Integrated Farming

Raising cattle is a way of life for many rural people in India and the application of cattle manure for fertilizing fishponds is an age-old practice. Consequently, the adaptation to fish-dairy integrated farming could be easily carried out by these farmers. For example, raw cattle dung could be used for fertilizing fish ponds and humus from the ponds, in turn, would make a good fertilizer for growing cattle fodder on dykes along the fish ponds. Recent attempts to estimate the optimum number of animals required to maximize productivity per unit of water when integrating small livestock and fish have revealed that five cows can produce 9500kg of milk along with 3,500 kg of fish, resulting in a net profit of Rs. 63,250/ha/yr.

Fish-Poultry Integrated Farming

Scientists have developed a simple and economically viable system for integrating fish culture and poultry farming system which reveals that, on an annual basis, 4,500 to 5,000 kg of fish, more than 70,000 eggs, and about 1250 kg of poultry meat can be produced by using 500 to 600 birds per ha of pond area. An annual net income of Rs.50000/per ha is possible with the prevailing market rates.

Fish-Duck Integrated Farming

As per ICAR experimental findings, for one ha pond area 200 numbers of ducks were reported to be sufficient for integration. In these experiments, 3700 kg of fish, 16,000 numbers of duck eggs and 500 kg of duck meat were obtained from one ha area in a year.

Fish-Pig Integrated Farming

Although the Chinese consider pigs as free fertilizer factories on hooves (FAO, 1977). Fish-pig integrated farming in Eastern India has yielded fish production ranging from 6,644.5 to 6,792.4 kg/ha/yr, without supplementary feeding and fertilizers. This compares very favourably to the 4250 to 4,448 kg/ha/yr, which has been achieved using intensive feeding and fertilizer applications. In addition, pork production ranging from 563.28 to 561.32 kg/yr was also realized. The cost of fish production was calculated to be Rs.2.48 to 2.4/kg, as compared to Rs.5.00/kg in conventional fish culture.

Integrated fish farming is a multi-commodity farming system in which fish culture is effectively combined with livestock rearing or crop cultivation. It is an entirely different farming system, which requires a new kind of technology.

Chapter 3
Fish Culture

Indian aquaculture is mainly carp-based wherein catla, rohu and mrigal are grown together under polyculture system or along with the three exotic carps, *viz.* silver carp, grass carp and common carp, as composite carp culture systems. These six species are selected because of their compatibility for habitat preference and food to utilize all the ecological niches of the culture system. While catla and silver carp are basically surface feeders showing preference for zooplankton and phytoplankton respectively, mrigal and common carp are omnivorous bottom feeders. Rohu is a column feeder and grass carp shows high preference for submerged vegetation. These carps which feed at the base of the food chain especially, phyto- and zooplankton, detritus and aquatic weeds, utilize natural productivity with greater output. However, among six species, the Indian major carps are slow growing as compared to their exotic counterparts. Thus, polyculture of three Indian carps usually results in lower production level than six-species composite culture system or the system with three exotic carps. These carps are also reared under mixed culture system incorporating freshwater prawn (*Macrobrachium rosenbergii, M. malcolmsonii*)

Carp culture is undertaken mostly in earthen ponds of varying dimensions. Several combinations of cultural practices have been evolved in the country to suit fish species, water resources, availability of fertilizer and feed resources. The standardized package of practice for carp polyculture comprises of:

1. Pond preparation,
2. Liming,
3. Fertilization,
4. Stocking,
5. Feeding and Feed management,
6. Water-quality management, and
7. Health management.

Grow-out culture of carps is carried out in earthen ponds ranging from 0.05 to 5.0 ha and 1-3 m in depth in different regions of the country; and ponds of 0.5-1.0 ha size with water depth of 1-2 m are considered ideal. While small and shallow stagnant ponds have several inherent problems, which adversely affect fish growth, the large and deep ponds have their own problems of management. Essentially, management practices in carp polyculture involves environmental and biological manipulation for obtaining higher levels of fish production, which can be broadly classified as pre-stocking, stocking and post-stocking operations.

Pre-stocking Management

Pre-stocking management includes the following practices:

1. Eradication of aquatic weeds
2. Eradication of predatory fishes
3. Eradication of weed fishes
4. Control of aquatic insects
5. Liming
6. Fertilization

Fertilization Practices

Pond fertilization aims at enhancement of autotrophic and heterotrophic production of ecosystem, which is stimulated by organic and inorganic fertilizers. In carp

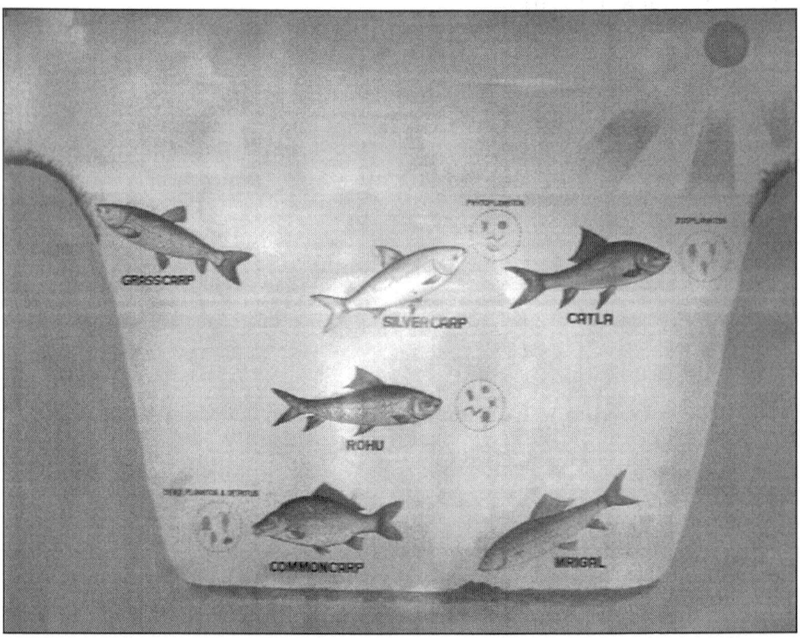

Figure 2: Composite Fish Culture System

culture, the usual practice involves application of both organic manures and nitrogenous and/or phosphatic fertilizers. Freshwater ponds of the country usually possess adequate amount of potassium required for natural productivity, thus potassic fertilizer are usually not supplied to culture ponds. Cattle-dung or poultry droppings are the most commonly used organic manures, often used in combination with urea and super-phosphate as inorganic nitrogen and phosphorus sources. The conventional dosage followed in carp culture practice in India usually ranges from 10 to 20 tonnes of cattle dung/ha/year or 4-8 tonnes of poultry manure/ha/year alone or in combination with urea @ 100 kg N/ha/year and superphosphate @ 50 kg P/ha/year. With regard to the total amount of organic manures, 25-30 per cent is generally applied as basal dose a fortnight before stocking and the remaining 70–75 per cent is applied in equal installments at fortnightly intervals. Beside cattle-dung and poultry droppings, other organic manures with high nitrogen and phosphorus contents such as pig manures, duck droppings and domestic sewage are also used depending on the availability.

Attention in recent years has been shifted towards usage of biofertilizers and bioprocessed organic materials for ensuring sustainability of aquaculture practices and to avoid possible environmental degradation. *Azolla*, a nitrogen-fixing aquatic fern, with high protein content (15-17 per cent), has been standardized as an ideal nitrogenous biofertilizer for aquaculture at 40 tonnes/ha/year; providing almost full complement of nutrients required for intensive carp culture (100 kg nitrogen, 25 kg phosphorus, 90 kg potassium and 1,500 kg organic matter). The detritus resulting from decomposition of *Azolla* applied in ponds serves as a trophic component of carps and prawns. The bioprocessed organic manures, biogas slurry, has been standardized as a manure in carp culture, @ 30-45 tonnes/ha/year, with distinct advantages in terms of lower oxygen consumption and faster nutrient liberation rates. The use of *Azolla* and biogas slurry has been found economical compared to traditional fertilization practices.

Stocking

The stocking density of carps for a culture pond is predetermined based on the targeted production, pond productivity and carrying capacity, species to be cultured, their feed-conversion efficiencies, size at stocking, growing period and level of management. Though fingerlings of more than 100 mm constitute best stocking material for grow-out culture, constraints in their availability often force farmers to resort to stocking of smaller fry. In intensive culture, a size of 50-100 g is preferred for stocking to realize higher survival and better growth. Generally a density of 5,000-10,000 fingerlings/ha is kept as a standard stocking rate in carp polyculture for a production target of 3-5 tonnes/ha/year. In seasonal ponds or in areas where water level becomes limited during summer, the stocking is reduced to 2,000-3,000 fingerlings/ha to obtain higher growth. Although major carps are expected to reach an average of 0.8-1.0 kg in the first year, the growth rate is invariably reduced at higher stocking densities. With provision of water exchange and aeration, higher targeted fish production levels of 10-15 tonnes/ha/year are achieved by resorting of stocking ponds at a density of 15,000-25,000/ha.

Selection of compatible species and stocking them at appropriate ratio suiting to culture environment is another important criteria in carp culture. Manipulation of species ratio is done for minimizing the interspecific and intraspecific competition for available food at various trophic levels in a pond. More than one species occupying different niches could be utilized in a pond for exploiting available natural food. Several combinations of Indian major carps alone or in combination with exotic carps have been experimented under different stocking densities, keeping growth as an index of performance. A combination of catla, silver carp, rohu, grass carp, mrigal and common carp fulfils the criteria of species, and has proven an ideal combination for freshwater carp culture in India. A proportion of 30-40 per cent surface feeders (silver carp and catla), 30-35 per cent mid-water feeders (rohu) and 30-40 per cent bottom feeders (common carp and mrigal) is commonly adopted depending on the productivity of the pond. When there is a possibility of providing aquatic vegetation as food for grass carp on a regular basis, the same may be included at 5-10 per cent of the total stocking. While ponds with higher organic matter can be stocked with higher proportion of bottom feeders, surface feeder may be kept more in newly excavated ponds. Similarly, *rohu* may be stocked at higher ratio in ponds with higher water depth.

Supplementary Feeding

The natural productivity of pond, irrespective of the level of its augmentation through fertilization, is not capable to sustain a higher level of fish biomass in semi-intensive grow-out. Due to the limitation in availability of natural fish food in pond at higher stocking density, the energy requirement for somatic growth can be met only through provision of supplementary feed. Over years, nutritional requirement of carps has been studied. Various ingredients of plant and animal origin, mixed at various combinations, have been evaluated in culture systems for production performance of the carp species. Among those, mixture of groundnut/mustard oilcake and rice-bran in the ratio of 1:1 has been commonly used in carp culture.

Several balanced feed formulations have been developed making up deficient essential amino acids, fatty acids, and incorporating vitamins and minerals. All these ingredients are blended together at required proportion for preparation of carp feed. The commercial production of supplementary feed is carried out in pellets of different diameter, so as to provide higher water stability, consumption and utilization by the fish. Under feeding depresses fish growth while overfeeding results in wastage of food, leading to deterioration of water quality. As feed cost constitutes more than 60 per cent of the recurring expenditure in carp culture feed management, is an important aspect for ensuring optimum growth of carps and limiting production cost, besides maintenance of aesthetic condition of ponds. The recommended practice is to provide feed at 3-5 per cent of body weight of stocking material initially and subsequently at sliding scale from 3 to 1 per cent.

The daily ration is adjusted according to the biomass estimated from monthly sampling and it is provided in two splits during morning and evening in feeding trays or perforated gunny bags suspended at regular intervals in pond. Grass carps are provided with aquatic vegetation like duckweeds, *viz. Lemna, Spirodela* and *Wolffia; Hydrilla, Najas* and *Ceratophyllum* at periodical intervals. To overcome shortage in supply of desired vegetation, land grasses, fodder, banana leaves and vegetable refuse can also be used.

Management of Soil-and-Water Quality

Management of soil-and-water quality in carp culture ponds is an integral part of the culture operation for a successful crop. A number of environmental problems are sometimes encountered in grow-out ponds. Some of these are inherently associated with the site characteristics like porosity, acidity and high organic-matter content of bottom soils, and some are encountered during cultural operations. Drying of pond bottom to crack between crops helps in aeration, which enhances microbial decomposition of soil organic matter. The porosity of the pond bottom is corrected through bentonite, lining with plastic sheet at 0.3-0.5 m depth and heavy doses of organic manures (cattle-dung at 10,000-15,000 kg/ha/year).

The water-quality parameters required for optimum growth of carps are pH 7.5-8.3, temperature 27-32°C, dissolved oxygen more than 4 mg/l, total alkalinity of 80-200 mg $CaCO_3$/l, Secchi disc visibility of 25-30 cm, total inorganic nitrogen 0.5-1.0 mg/l and phosphorus 0.2-0.3 mg/l. Variations in these water parameters occur in culture pond and need periodic correction/management measures. The low water pH is corrected through intermittent application of lime materials. Lime helps in improving alkalinity, hardness, controlling turbidity and reducing H_2S build-up. Dolomite is particularly useful when augmentation of phytoplankton growth is required.

Agricultural gypsum ($CaSO_4$) is applied to correct alkaline pH. It is also applied to increase total hardness without affecting alkalinity. Aeration, a proven method for improving pond dissolved oxygen availability, also helps in mineralization process reducing organic load. Water exchange helps to a great extent in reducing metabolic load. Addition of water into pond, particularly during winter, helps in improving temperature regime and prevent temperature stratification. Turbidity with suspended soil particles can be controlled by cattle manure (500-1,000 kg/ha), gypsum (250-500 kg/ha) or alum (25-50 kg/ha). The ammonia load in pond can be reduced through encouraging healthy growth of phytoplankton or aeration. The Hydrogen Sulphide (H_2S) build up in pond can be subsided through frequent water exchange.

Health Management

A balanced relationship among the host, pathogen and the environment is a prerequisite for optimum growth and health of any organism. Any imbalance in this relationship leads to diseased condition of the host, *i.e.* animal. Fish is a cold-blooded animal and more prone to environmental deterioration. In the event of a disease outbreak, it affects fishes in groups or the whole population. Further, treating a diseased fish is relatively difficult compared to terrestrial animals, as individual treatment is not possible in former, thereby requiring mass treatment of population or environment. Therefore, prevention is always a preferred method in aquaculture to control disease outbreak than curing disease. In recent years, due to increase in intensity of culture, disease management and health care has become an important aspect of management for preventing sudden outbreak of epizootics that occur due to environmental deterioration, improper feeding and overcrowding.

Underfeeding leads to malnutrition, resulting in retardation of growth and lowered disease resistance. Many of the fishes carry small number of pathogens like bacteria, viruses, fungi and parasites, either as chronic low-grade infections or serving as carriers. Such pathogens are also present in deteriorated pond-water and sediment and their population increases to a very high level when pond is heavily stocked. Weak disease resistance of the fish makes them susceptible to attack from these pathogens.

The best way to avoid disease outbreak in pond during culture is through taking preventive measures which are ensured by proper management of the soil-and-water quality, following proper feeding schedule, use of balanced feed, periodic sampling for health check, and minimizing outside influence on the pond.

Harvesting and Marketing

Generally carps are harvested after a grow-out period of one year during which it reaches marketable size of 0.8-1.0 kg. However, these carps are even marketed in smaller sizes of over 300 grams (except silver carp; since the marketable size is over 1 kg). In multi-harvesting system, the fishes attaining market size are periodically harvested from the pond and the smaller ones are released back for further growth. While intermittent harvestings are done with dragnet of suitable mesh size, final harvesting is usually done by complete draining of pond.

Carps produced from culture ponds are mostly sold in local market, either in live or dead condition. The Indian

Figure 3: Harvesting of Fishes

major carps are also transported to adjacent deficit areas as well as to distant places, even 2,000-3,000 km away from the production site, in insulated vans with ice. Fresh fish fetches about one-and-half-times higher market price than iced ones. When sold in live condition, the carps command still higher sale value of over two times compared to iced ones. The price in the domestic market is influenced by demand and supply.

Chapter 4

Cattle-Fish Integration

Fish farming using cattle manure is one of the common practices all over the world. Among all livestock excreta, cattle excreta is the most abundant one in terms of availability. A healthy cow weighing about 400-450 kg excretes over 4,000-5,000 kg dung, 3,500-4,000 litre urine on an annual basis. The manure particles of cattle sink at 2-6 cm/minute as against 4.3 cm/minute of pig manure. This provides sufficient time for fish to consume edible portion available in dung. Furthermore, biological oxygen demand (BOD) of cattle manure is lower than that of other livestock manures since they are already decomposed by micro-organisms present in rumen. Manuring of fish ponds with cowdung (which is rich in nutrients) results in increase of natural food organisms-detritus and bacteria. The faeces and urine are extremely beneficial for filter feeding and omnivorous fishes such as catla and silver carp. A unit of 5-6 cows can provide adequate manure for 1 hectare of

Figure 4: Cattle-cum-Fish Culture

pond. In addition to 9,000 kg of milk, about 3,000-4,000 kg fish/ha/year can also be harvested with such integration. Farmers gets returns from both milk and fish which in turn increases his household security and subsequently reduces working expenditure.

The cow-sheds should be built close to fish-pond to simplify handling of cow-manure. The excreta and the urine is collected separately or can be flushed directly into fish-pond. This method saves time and labour. Grown-up cow requires about 7,000-8,000 kg of green grass annually. The leftover grasses which are about 2,500 kg, are utilized by grass carp. The concentrate feed wasted by cows, comprising of grains and oilcakes which consists of grains is also utilized by fish. In place of raw cowdung, biogas slurry could also be used with equally good production. Digestion of raw cowdung in biogas plants help in keeping pond environment

Cattle and fish integration is a common model of integration. The urine of cattle, daily washings and manure can be disposed at hand resulting in saving energy, labour and money. In this integration, the fish farmers get sizeable money by way of supply of both fish and milk to the market. Cattle provides organic manure to fish ponds. The cattle dung is rich in nutrients and fine in texture due to repeated digestion in the four compartmental stomach of cattle. Hence the dung suspends longer in water.

The suspend ability not only enables fish to get more feeds, but also reduces oxygen consumption caused by manures and also avoids the formation of harmful gases. The Biochemical Oxygen Demand (BOD) of cattle dung is relatively less than other livestock manures, since they are already decomposed by microorganisms in the cattle stomach itself. Hence there is very limited risk for the fish ponds using cattle manure. The nutritional elements of cattle dung and urine in fresh matter are as follows.

(*i*) Organic matter (per cent)

Cattle dung: 14.0

Cattle urine: 2.3

(*ii*) Inorganic matter (per cent)	N	P	K
Cattle dung	0.30	0.2	0.1
Cattle urine	1.0	0.1	0.4

Five to six numbers of cattle (cow or buffalo) are sufficient for 1 hectare water spread area of fish ponds. The cattle shed constructed on the pond bund should have a concrete flooring so as to drain the daily washings along with cattle dung into the pond. The fodder raised on the pond bunds can be used for feeding the cattle. In this type of integrated fish farming, inputs such as fertilizers and supplementary feed for fish are not required and hence there is considerable reduction in the cost of fish production.

The cattle waste like urine and washings when stagnated in the vicinity of cattle shed, is an ideal breeding ground for mosquitoes and disease spreading organisms. But if these wastes are utilized for fertilizing the fish ponds, fish can be economically cultured and at the same time spreading of disease causing organisms are also prevented.

Cattle Housing

Cattle may be raised either for beef or milk production. The type of cattle housing depends on herd size, climatic conditions, labour availability, amount of capital investment, degree of mechanization and owner's preference. Many factors should be considered while constructing a cattle shed. The most important considerations are:

Provide adequate comfort to the cows so that maximum milk production is obtained,

Minimize injuries such as swollen hocks, lameness by way of non-slippery floors.

Minimize exposure to environmental stresses, considering climate of the geographical region involved.

Follow a feeding system that allows sufficient feed for all cows housed with minimal labour.

Fulfill sanitary requirements and keep disease incidence to a minimum.

Choose a milking system that minimizes labour while producing higher milk yields from the dairy cows.

Choose a waste management system that will provide maximum economic return.

The choice of a disposal system is based on:

Size and type of operation

Availability of land and cropping practices,

Population density around the location.

Climatic conditions, and

Type of housing arrangement.

Care in correct operation and management of disposal systems leads to cleaner cows, lower possibility of disease outbreaks and less problems related to odour.

Five distinct housing types are employed in raising dairy cattle. They are:

(*a*) Stantion or Tie Stall System

Cows are confined to individual stalls by a stantion or some type of restraining device. A paved area is included where the cattle are released for exercise. The time the cows spend in stalls depends on the pasture availability and season. Milking is done in the stalls with portable can milkers, a pipeline milking system, or in a milking parlour. The advantages of this type of housing are, more efficient feeding of individual cows, greater attention for individual cows, better display of cows, and environmental control.

(*b*) Free Stall Housing

It consists of four distinct parts known as the feeding area, stall area, milking parlour and exercise area. Cows are allowed free movement between feeding and stall areas. They are pastured up to 80 per cent of the time depending on climate and land availability. Milking is performed twice a day in the parlour. The housing can be made either with cheaper pole-type construction or by costly construction methods where greater environmental control is needed. The advantages of this system are less injuries, cleaner and happier cows, less usage of bedding materials, more efficient use of labour and lower floor space requirements per cow.

(*c*) Loose Housing

It is the oldest system for dairy cattle management. A loafing area or barn is provided which is heavily bedded with a manure pack. It is generally open to the environment with a paved exercise lot connecting the feeding and resting areas. The advantages are lower labour and building costs due to the adaptability of more mechanized feeding technology and lower construction costs of pole-type facilities.

(*d*) Partially Housed or Non-Protected Cow yard System

It is used where minimal environment protection is needed. It employs an open or partially sheltered lot where cattle are fenced. A sunshade is built on the lot for partial protection of the cows from solar radiation and rain. Cattle are fed in feed bunks directly on the lots, which are paved, and semi paved. Milking in this system is done in a milking parlour twice a day.

(*e*) Pastures

The method of pasturing is a good way to get full use out of forage materials on fields that are not suitable for harvesting. Milking cows must be returned to the milk house or milking parlour twice a day. If the herd is of considerable size or the fields are distant, the labour needed would be prohibitive. Pasturing is not ideal for calves, which need greater attention, and bulls that are likely to knock down fences. Considerable savings are realized through lower feed costs. If forage materials are not of high quality additional concentrate feeding must be undertaken or else milk production will be reduced. Shade should be provided by employing an open shelter or wooded area for protection from heat and solar radiation, since they have a deleterious effect on the cows' milk yield. The feeding practices of dairy cattle must be managed properly and efficiently to insure a steady production of quality milk. Feeding costs account for 40 to 65 per cent of the total cost for milk production. Underfeeding to reduce feed costs is undesirable. Lower feeding rates will reduce the milk production per cow and net income; this is because all cows require the same basic maintenance ration. All additional feeding above this maintenance level is converted to milk resulting in lowest feed costs per pound of milk produced for the highest yielding cows. To maximize production for each cow, the feeding program must ensure that the individual cow receive adequate rations. Dairy cows are fed with a mixture of roughages, concentrates and supplements in quantities that will fulfill their nutritional requirements. These requirements are determined by animal weight, productivity level and butter fat percentage in the milk. The cattle are either fed in confined barn area or are pastured receiving supplemental feed in the field or milking area. Increased concentrate feeding, has contributed to doubling of the output of the milk produced per cow. The quality of concentrates fed has an important effect on feed and proper nutrition. A successful feeding operation must satisfy the following criteria:

1. Keeping calves alive and healthy
2. Producing a growth rate that will allow calves to breed at about 15 months of age.
3. Producing a good desirable stock, and
4. Should be economically feasible.

It is essential that the newborn calf is fed with colostrum (first milk after birth) during the first 3 or 4 days. This milk is rich in proteins, nutrients and antibodies that are needed by the calf to prevent against diseases. It should be fed frequently in small amounts directly from the mother (cow) or by artificial means through feeding bottle after being milked from the mother. Once the calves are weaned (separated from mother), they should receive a diet high in concentrate materials supplemented with good quality roughage. Concentrates must be fed until about six months of age in order to allow time for rumen development accompanied by more efficient roughage digestion. If calves are pastured during this period, they must receive these materials in the field. Water must be available at all times because of the high water requirements for maintenance and growth.

Chapter 5
Integrated Pig-Fish Farming

Integrated pig-fish culture is not a new concept; it has been practiced for many years in various parts of Asia. It is the most prevailing integrated fish farm model practiced throughout China (Yang, 1994). Pigs have a special significance in their agricultural economy as they are considered 'costless fertilizer factories, moving on hooves', and pig manure forms the main source of home-made fertilizer used by farmer to meet the needs of crop farming, as well as fish farming. Apart from ducks, pigs are the farm animals most commonly raised in combination with fish (Wohlfarth and Schroeder, 1979). Pig-fish rearing in backyard is most successful in Vietnam and Philippines. This farming practice has been well studied and analyzed. Both traditional and modern intensive pig-rearing facilities have been successfully adapted for pig-fish culture (Little and Muir, 1987). Fish farmer can reduce the production cost of fish to an extent of 60 per cent by reducing the feed and fertilizers with pig manure. The following advantages could be obtained through fish cum pig integration.

(*i*) Pig-fish culture maximizes land use by integrating two farm enterprises in the same area.

(*ii*) The left over residues of kitchen, aquatic plants, agriculture by-products and wastes are used as feed for pigs and pig excreta are in turn used as organic manure for fish ponds.

(*iii*) Per unit area animal protein production in pig-fish integration is very high when compared to other integrated fish farming systems,

(*iv*) Aquatic weed if any in the pond are utilized as roughage to pigs.

(*v*) Helps in environmental sanitation.

(*vi*) The pig dung acts as an excellent fertilizer for ponds and it also raises the biological productivity of the pond thus increasing fish production.

(*vii*) Some of the fishes feed directly on the pig excreta which contains 70 per cent digestible food for the fish.

(*viii*) No supplementary feed is required for the fish culture, which normally accounts for 60 per cent of the total input cost in conventional fish culture.

(*ix*) Pond water can be used for cleaning the pig sheds and for bathing pigs.

(*x*) The fishpond serves as a sanitary disposal place for pig wastes (*xi*) The pond dykes provide space for erection of pig housing units

(*xii*) Pig-fish integration provides additional income and also acts as a cheap source of animal protein.

In India, pig-fish farming has a special significance as it can improve the socio-economic status of many of the weaker rural communities especially the tribal communities which traditionally rear pigs. The procedure for pig/fish rearing can follow a number of patterns. In its simplest form it is the fattening of piglets (2-3 months old) to a marketable size (60-100 kg) over a feeding period of 5-6 months. This permits two production cycles per year, each corresponding to one fish culture cycle. Feeding occurred at a rate of approximately 3.5-7 per cent body weight per day, the pigs being then marketed at a mean weight of approximately 80-150 kg. Within one growing cycle of pigs, two production cycles (2x90 days) of tilapia (*Oreochromis niloticus*) can be achieved. In India too, efforts have been made to popularize system and in this connection several exotic breeds of pigs have been introduced to augment production of pigs.

Figure 5: Pig-cum-Fish Culture

Breeding Stock

An important factor in pig-fish culture is the breed of livestock used. In India, Large White Yorkshire and Landrace piglets are used (Jhingran and Sharma, 1980), while a large white-landrace cross is predominantly used in the Philippines (Cruz and Shehadeh, 1980). In Africa local breeds of pigs are adapted to pasture and not commonly kept in enclosures.

Variety of Pigs Used for Integration

The following varieties could be used for integration with fish culture.

They are (*i*) large and middle white Yorkshire, (*ii*) Berkshire, (*iii*) Hampshire, (*iv*) Tam worth, (*v*) Saddle back and (*vi*) Landrace. Of these, large white Yorkshire and landrace are being widely used; the others are being used in some small pockets.

Large White Yorkshire

This breed is native of England and is reared extensively in India, The body colour is solid white with occasional black pigmental spots. Other features are erect ears, snout of medium length and dished face. It is a prolific breeder.

Landrace

The native of this breed is Denmark and is white in colour with black skin spot freckles. The other features are long body, large dropping ears and long snout.

Pig Housing

Basic considerations in providing proper structures are necessary for housing pigs, particularly for exotic varieties and they should be provided proper hygienic conditions.

The housing system may be indoor or open-air system (extensive). The pigs house is collectively called as pigsty. Pigsty has one or more pens. In each pen, one or more pigs may be kept. In indoor system, the pigsty is usually constructed in one corner of the dyke of the pond.

The pens constructed for accommodating pigs should have adequate floor space area and facilities for feeding and providing drinking water. The foundation of the pigsty should be at least 4 times wider than structural walls and deep enough. The floor should be cement concrete so as to prevent the subsoil water from rising up the building by capillary action and easy to clean. The height of the pigsty should be 2 to 2.5m from floor level. The top portion of the pigsty is covered with thatched roof, RCC flat type or gabled roof. The gabled roof may be asbestos, cement, steel sheet or clay tiles. The side of the pigsty is covered with iron mesh frame. Proper drainage should be provided and minimum width of drain should be 0.25 m with a slope of 2.5 cm in every10 m. Over heating should not be allowed to occur in the pigsty.

Pigs need clean housing which should provide adequate protection from adverse climates. A run or courtyard adjacent to pen is essential. The size of the pigsty depends on the number of pigs required to provide sufficient manure for water-body. Each pig should be provided with a floor space of 1-1.5 m^2. The pigsties are build mostly at the pond sites and at times even over pond. The washing of the pens containing dung and urine is either directly channelized into ponds or partially decomposed before applying.

Alternatively the sty may be constructed over the fish pond. Typically, the structure is supported on wooden slits over the pond and provided with a lattice type of floor which permits the excreta and uneaten food to fall directly into the pond water without requiring the regular washing out and sluicing of the traditional concrete based housing. This type of construction may also have some cost advantages although this is largely dependent on the cost and availability of wood compared to concrete.

Feeding of Pigs

Feed plays an important role in successful pig production. The quality of the ration determines to a great extent the rate of growth in young pigs, general resistance to diseases and parasites, regularity of breeding and the vigour of litter. Among the livestock pigs grow more rapidly and hence suffer more from nutritional deficiencies than the ruminants. They are also fed with chopped grasses once in a week to prevent mineral deficiency. A complete diet consists of protein (more than 20 per cent), carbohydrates, fats, minerals, vitamins and clean water.

The Table 1 gives the various requirements in the formation creep, grower and finisher rations for pigs.

The Table 2 shows the amount of feed (feeding rate) to be given to pig/day:

Pig should be fed twice a day. To reduce the quantity of concentrate ration and also to reduce cost, spoiled vegetables, especially boiled potatoes can be mixed with pig mash and fed Traditionally, in some areas, pigs are fed with spill feed or waste food from households or restaurants rather than balanced, concentrate feeds. The pigs in integrated farms traditionally depend on feedstuffs produced on the farm. Aquatic plants such as *azolla*, duck weed, water hyacinth (*Eichhornia crassipes*), *Ipomoea*,

Table 1: Requirements of Ration for Different Stages of Pig

Sl.No.	Components	Creep Feed (up to Weaning)	Grower Ratio (20-40kg)	Finisher Ratio
1.	Protein Supplements (per cent)			
	(*i*) Oil cakes	16–18	14–16	13–14
	(*ii*) Animal protein	8.10	4	2
2.	Grains (maize, millets, Sorghum) (per cent)	60–65	50–55	40–50
3.	Wheat bran (or) Rice bran (per cent)	5	10	20
4.	Mineral-mixture (per cent)	0.5	0.5	0.5
5.	Antibiotic supplement (mg)	40	20	10

Table 2: Feeding Rate for Pigs

Weight of Pigs (kg)	Daily Consumption of Feed (kg/pig)
7.5	0.7
10–12	1.2
20–25	1.3
30–35	1.5
45–50	1.6
70–80	1.7
90–100	1.8
120–130	2.0

Wolffia, Lemna and *Pistia stratiotes* grown on feeder channels and irrigation ditches associated with the pond farms and are used for feeding the pigs. In addition, the foliage of several terrestrial plants such as vegetables, corn, rice and leguminous plants are utilized as feedstuffs for pigs. These plant materials are generally mixed with rice bran, bananas, coconut meal, soybean wastes, fishmeal etc. for feeding to the pigs.

Pigs Excreta as Manure

Washings of pigsties containing urine, dung and spill over are channeled into pond. Alternatively, pig-dung collected from pig-houses is applied to pond every morning after sunrise. Each fully grown pig voids between 500 and 600 kg dung in a year and excreta released by 40-45 pigs is adequate to fertilize 1 ha water area under polyculture of fish during the year. Lime is applied at 250-350 kg/ha/ year depending upon soil and water condition. Half of the quantity is applied before stocking fish and the rest in 2-4 installments as and when necessary. This pattern of liming the pond keeps water in the pH range of 7. Growth of pigs depends upon many factors including their breed and strain but good management contributes considerably to optimum production. Pigs attain the marketing weight (slaughter–60-70 kg) within 6 months in case of pigs raised for pork production. The pigs reared for breeding deliver 6-12 piglets in every litter. Their age at first maturity ranges form 6 to 8 months. Fish attains marketable size in a year during which period more than one generation of pigs reach marketing weight.

Harvesting

Due to abundance of natural food in integrated fish-cum-pig pond, some fishes attain marketable size within a few months instead of a year. Keeping in view the size attained, prevailing market rate, demand of fish in the local market, partial harvesting of table-sized fish is done. After harvesting partially, stock in the pond is replenished with same number of fingerlings as the fishes removed, depending upon the availability of fish-seed. Final harvesting is done after 12 months of rearing. Fish yield ranging from 3,000 to 4,000 kg/ha/year is generally obtained.

Table 3: Guidelines for Normal Reproduction of Pigs

Age at puberty	6 to 7 months
Breeding age of gilts	10 to 12 months
Breeding weight of gilts	90 to 100 kg
Breeding age of boar	18 to 24 months
Number of sows per boar	10
Heat cycle	19-23 days (Ave.21)
Heat period	2-3 days
Mating time	Gilts-first day of heat and second day of onset of heat in sows
Number of services per conception	Two at interval of 12 to 14 hours
Gestation period	112 to 114 days (3 months, 3 weeks, 3 days)
Suckling period	56-60 days
Average litter size at birth	10 to 14
Average litter size at weaning	8 to 10
Rest period	45 days
Occurrence of heat after weaning	2 to 10 days
Period of mating	15 days after weaning
Volume of ejaculate	200 c.c.
Average number of sperms/cumm.	100,000
Average age to castrate pigs	4 to 8 weeks
Marketing age of fattening pigs	6 months
Market weight at 6 months	70 to 75 kg
Farrowing interval	7 to 7½ months
Years-sows are known to breed	8 to 10 years

Characteristics of Pig Manure

Pig manure is fine in texture. Pig manure contains higher phosphorus levels than cow dung. It contains major inorganic nutrient components (N,P,K) in addition to such trace elements such as Ca, Cu, Zn, Fe and Mg. The major portion of pig urine is nitrogen in the form of urea and decomposes easily. It has been reported that a pig excretes about 1000 kg of faeces and 1200kg of urine in the fish culturing period of 8 months *i.e.*, from piglet to an adult size sow. Pig excretes about 10-20 per cent of its body weight.

Nutritional Elements (Chemical Composition) of Pig Manure and Urine

The chemical composition of pig manure and urine in fresh matter is given in Table 4.

Table 4: Chemical Composition of Pig Manure and Urine

Sl.No.	*Components*	*Faeces (%)*	*Urine (%)*
1.	Moisture	70-77	90-95
2.	Dry matter	23-30	5-10
3.	Organic matter	15	2.54.
4.	Inorganic matter		
	(*a*) Total carbon	2.72–4.00	–
	(*b*) Total nitrogen	0.60	0.40
	(*c*) Ammonia nitrogen	0.24–0.27	–
	(*d*) Phosphorus	0.50	0.10
	(*e*) Potassium	0.40	0.70
5.	C/N rartio	14:1	–
6.	pH	6.6-6.8	7.6

Application of Pig Manure

Pig manure is used as fertilizer in fishponds in raw or

fermented form. The fermentation of pig manure is done by composting.

Technique of Composting

1. Pig manure is collected and placed in composting pits located in one corner of the farm. Compost pits are usually circular having 2.5m bottom diameter, 1.5m deep and 3m top diameters.
2. Pit is filled by layering river silt (97.5t)/rice straw (0.15t) mixture, pig manure (1 ton) and aquatic plants or green manure crops (0.75t) each in 15cm layers.
3. The top is covered with mud and a water column of 3-4cm depth is kept at the hollowed surface in order to create anaerobic conditions. This minimizes the nitrogen losses.
4. The contents of the pit are turned over after 1,2 and 2½ month after which the compost is ready.

Each pit produces about 8 tonnes of compost adequate to fertilize 0.1 hectare of area of cropland. Compost is applied to fishponds ranging from 5 to 10t/ha/depending upon the type of soil.

Fresh, Untreated Pig Manure

The application of fresh and untreated pig manure to fish ponds are possible where pigsties are situated over the ponds. The application of fresh and untreated manure to fish ponds has given high fish yields. Hence, daily fresh pig manure application to a pond gives maximum yields than pig compost.

The application of pig dung in ponds provides a nutrient base for dense bloom of phytoplankton, which in turn forms a base for intense zooplankton development. Zooplankton have an additional food source in bacteria which thrive on organic fraction of added pig dung. This calls for stocking filter feeding phytoplanktophagous and zooplanktophagous fishes. In addition to above, production of detritus at pond bottom also starts to provide substrate for colonization of micro-organisms and other benthic fauna, especially chironomid larvae. Thus, while stocking, emphasis must be given on bottom feeders. Polyculture of Indian major carps, catla (*Catla catla*), rohu (*Labeo rohita*), mrigal (*Cirrhinus mrigala*) and exotic carps, silver carp (*Hypophthalmichthys molitrix*), grass carp (*Ctenopharyngodon idella*) and common carp (*Cyprinus carpio*) is undertaken in fish-cum-pig-farming ponds. Pond is stocked only after the pond-water is properly detoxified. Stocking rates may vary from 8,000 to 8,500 fingerlings/ha, and a species ratio of 40 per cent surface feeders, 20 per cent column feeders, 20-30 per cent bottom feeders and 10-20 per cent macro-vegetation feeders (grass carp) is preferred for high fish yields. In case of pigs it takes 8-9 months in winter and 9 to 11 months in summer to reach a body weight of 70-80 kg. At weaning time, the piglets weigh about 15-16 kg. Under best management practices, pigs have reached 70-80kg body weight in 6 months time.

Pigs are resource intensive as they need a concentrate based diet to grow and produce quality wastes for fishpond fertilization. This technology is more applicable in farming systems where pigs are penned, not in small – scale rural farms which typically permit pigs to roam and scavenge for their food to avoid more investment in housing and feeding.

Considerable care and water management skills are required to prevent pollution of the water and mortality of the fish stock by pig manure. The common practice of direct washing of the wastes into the ponds should be avoided. The wastes should be conveyed to a specially built tank, where sedimentation and fermentation of the manure can take place. At regular intervals, the supernatant liquid from the tank should be allowed to flow into the ponds. The sludge that remains should be removed for fertilizing agricultural crops. Thus, loading of decomposable organic matter in the ponds can be reduced.

Consumers may be reluctant to buy fishes produced in manure-loaded ponds. Occasionally fish from ponds overloaded with pig manure have a "muddy" or off-flavor taste. It can be avoided by stopping manure loading to the pond a few days before harvesting the fish and transferring the harvested fish to a net enclosure installed in a clear pond at least 4-6 hours (preferably several days) prior to selling or eating them.

The system of pig-fish culture is quite significant in improving the socio-economic status of poor farmers.

Chapter 6
Rabbit-Fish Integration

Integration of rabbit production with fish culture has not received much attention, but seems to have considerable potential. Large-scale, intensive production of rabbits is not common, although units of up to 10000 breeding females have been reported in Europe. In many ways rabbit, if locally marketable, is the ideal animal for integration with small-scale fish culture. Confinement can be in a similar manner to that used for chickens, over the pond in simple bamboo or wood cages. Pilot-scale rabbit production in North-East Thailand suggests such husbandry practices may also reduce juvenile mortality and improve male fertility (often a problem in hot climates). Disease problems that are common in large-scale intensive production might also be reduced in small-scale ventures.

Rabbit manure may have greater value as a direct food for fish compared to other livestock wastes. Certainly, the pellet size and semi-floating nature of rabbit manure encourages enthusiastic and direct ingestion by fish. The proportion of true protein, compared to non-protein nitrogen is high in rabbit manure. Like wise energy value is also high when compared to dried poultry waste, broiler and cattle wastes.

Rabbit was considered, as a pet animal by the common man, and as an experimental animal by professionals during the last few decades, but currently it has emerged as an alternate meat source and it play an important role as a non-conventional meat animal. Rabbit meat has been regarded as a safe diet for health-conscious meat consumers because of its low fat content in comparison to chicken, mutton, beef and pork. Small body size, high reproduction rate, potential for year-round meat production and ability to utilize non-competitive feed, etc. are some of the qualities which make the rabbit an ideal candidate animal for meat production.

Approximately 60 individual breeds and varieties of rabbits are recognized all-over the world. Some of the important meat breeds are New Zealand White, Soviet Chinchilla, Grey giant and White Giant, while the wool type breeds are Russian Angora and German Angora.

Rabbits are reared in cage, hutch and floor systems. Cage system is used when rabbits are reared in semi-commercial and commercial scale. Hutch system is generally used for breeding and maturity purpose. Rabbit is a monogastic animal but presence of microflora in the hind gut (caecum) and habit of coprophagy makes it capable of consuming a variety of feed. Availability of water to rabbits should be assured. Rabbit excreta contain 50 per cent organic matter, 2.0 per cent nitrogen, 1.33 per cent phosphate and

1.2 per cent potassium. Rabbit excreta is low in moisture and high in nitrogen content and it can be very well used for sustained planktons production and hence rabbits can be efficiently integrated with fish-farming.

Housing System

Sheds

Atmospheric temperature, ventilation, humidity and lighting are the important points to be taken care of, while constructing the rabbit house. A shed with 4.2 m central height, 3.0 m side height and with gable roof is most ideal one. Locally available materials may be used in constructing the sheds. The commonly used materials are wood, bricks, asbestos sheets and wire nettings.

The ventilation system is to be so arranged as to remove stale air from the shed and replace it with fresh air. Though rabbits can thrive at temperatures between –2°C to 30°C, the ideal temperature is 15°C to 20°C. Higher ambient temperature reduces breeding efficiency, litter size, litter weight and reduces feed intake. The ideal humidity for indoor rabbitary is about 75 per cent at a temperature of 16°C. The relative humidity should not be higher than that of outside air by more than 5 per cent and the temperature should not differ from outside by 10°C.

Light can be provided by installing one 100-Watt bulb 40-Watt fluorescent tube 2 m above the floor at a distance of 3 m each, with an effective light of 16 hours in a day. *Kutcha* floor is preferred because it absorbs the urine and water and also reduces the ammonia odour generally prevailing inside the shed.

Cages

Cages are generally made up of wire netting and can be put either in a single row or in tier system. The materials must be galvanized to prevent corrosive effect of urine. Cage size should be 0.75 m x 0.75 m with 0.65 m height. A 1.25 x 1.25 cm mesh should be used for the floor and up to 15 cm height to the side of the cage. The remaining portion may be of 2.5 x 2.5 cm mesh. After weaning, colony pens may be used to rear rabbits till marketing.

Nest Box

The size of the nest should be around 1.8 x 0.9 x 0.9 m. After 30 cm length there should be tapering cut so that the height of the end will be 15 cm. The nest box may contain a bedding of 5 to 8 cm of wool, wood scrapings or coconut fibres. A few holes should be provided in the floor to drain out the urine.

Feeding Rabbits

The feeders are attached to the cages so that feed can be added from outside. The feeder should be 5-8 cm high so as to minimize contamination of feed by faeces, urine or water. An adequate supply of fresh and clean water is essential for good rabbit production. Earthen and aluminium bowls are cheaper and best suited for rabbits.

Feeding and Watering

Regular time-table for feeding is advisable so that rabbits do not feel any stress due to changes in their daily routine. The concentrate can be given in the morning and greens in the afternoon. Pellet feed of 3-4 mm diameter and 10-15 mm length are preferable than finely grounded feed.

It is preferable to make water available round the clock, particularly to lactating does. Concentrate feed includes grain such as Oats, Barley, Maize etc. (energy sources) and protein supplements namely of soyabean meal, groundnut cake, seasame meal etc. Table wastes, kitchen wastes, excluding meat, fat and spoiled feeds can be used to feed rabbits. Common salt is added to about 0.5 per cent of whole diet.

Table 5: Nutrient Requirement of Rabbit of Different Physiological Status

Nutrients	*Maintenance*	*Growth/Gestation Lactation*
Digestible Energy (k. Cal)	2100	2500
Crude Fibre (per cent)	14	10-12
Fat (per cent)	2	2
Crude Protein (per cent)	17	21

Green feeds include cabbage, oats etc., Sesbania leaves are best suited to feed rabbits in summer. Lucerne and stylosanthes hemata are considered as best leguminous fodder (in the form of pasture) for feeding rabbits.

Management

Good management is closely related to handling of animals, feeding, breeding, care of young ones, identification, disease control, shearing, slaughter, sanitation etc.

Handling

Rabbits should be handled gently and carefully so that it is not frightened. Rabbits are generally lifted by grasping the skin over the shoulders. They must never be lifted by ears, but ears should be included along with skin over the shoulders. One hand should be placed below the hindquarters for additional support.

Reproduction Management

The approximate age of first service is about 5-6 months age. Mating is done either early in the morning or in the evening. The doe is taken to the buck cage and never vice versa. After successful mating the male usually produces a typical cry and falls down to one side of the doe. The gestation period is around 30 to 32 days. The doe can be rebred even after one week of kindling in intensive system of breeding but normally is bred four weeks after kindling. The nest box is kept in the cage around 25th day of pregnancy to facilitate the doe for preparing bedding for the new born.

During summer season when temperature rises above 36°C temporary sterility sets in. This can be prevented to some extend by providing cool conditions.

Kindling

The youngones are born hairless with their eyes closed; usually 6 to 12 kids born in a single kindling. They open their eyes around 10-14 days. The nest box is removed after 3rd week of kindling when the kids start coming outside the box and try to consume some feed/hay. The doe should not be disturbed frequently. During the period the doe should be fed *adlibitum*. Adequate feeding and watering is essential to prevent the doe from crushing and eating the kids. Rabbits nurse their youngones only once daily, usually at night or in the early morning.

Weaning

Weaning is a stressful period and requires careful handling. The weaning is done between 4th and 6th week. The earlier the better. The doe should be removed from the cage and the litter allowed to remain behind together for next 3 to 4 weeks. Sex of the young rabbit can be determined one week after kindling by an experienced person.

Sanitation

Strict sanitation practices helps in disease control. Building should be well ventilated and adequately lighted. Manure should be removed regularly in weekly intervals to reduce ammonia and moisture level in the sheds.

Metal surfaces should be burnt by blow lamp and other materials may be disinfected with phenyl 1.2 per cent or Formaline 1 per cent or sodium hydroxide 2 to 4 per cent or Ammonia 10 per cent solution. Buck cages should be disinfected at regular intervals of 3 days during the mating period. The nest should be thoroughly disinfected before being put to use. During kindling period the cage should be kept free from dung material so that kids do not pick up injection.

Chapter 7
Goat-Fish Integration

The small size of these animals, their early maturity and low capital investment per head are particularly advantageous for small-scale farmers when compared with the large ruminants. Goats and sheep are found to play a significant role in many types of small farm system in Asia, Africa and Latin America. In Asia, over 60 per cent of the sheep and goat population is raised on farms up to five hectares in size and these ruminants therefore could be used to improve the productivity of small-scale fish culture. The importance of sheep and goats in many developing countries is now being recognized but this is not yet reflected in any widespread integration with fish culture. Perhaps this is not surprising since sheep and goats are that stocked on very extensive pasture in poorly irrigated and dry regions where fish culture is more problematic and has no tradition.

Goat is considered as poor man's cow and its farming is an age-old practice for meat, milk, fibre and manure.

Integration with fish has not been a common practice although it is practiced in some South-east Asian countries like Indonesia, etc. Indian has 25 per cent of the world's goat population.

There are 13 well-known breeds of goat apart from local nondescripts, distributed in Himalayan region (Chamba, Gaddi, Kashmiri, Pashmina, Chegu), Northern region (Jamunapari, Beetal, Barbari), Central region (Marwari, Zelwadi, Berari, Kathiawari, Sirohi, Jhakrana), Southern region (Surti, Deccani, Osmanabadi, Tellichery) and Eastern region (Bengal, Gangam Assam hilly breed). These breeds are reared for fibre, meat, and milk and are categorized as follows:

Goat Breeds

Fibre	*Meat*	*Milk/Meat*
Himalayan,	Bengal,	Jamanapari,
Chegu	Deccani,	Beetal,
	Osmanabadi,	Barbari,
	Jhakrana,	Marwari,
	Sirohi	Mehsana,
		Surti,
	Malabari (Tellichery)	

Housing

Goat-housing should be dry, comfortable, safe and protected from excessive heat. Adequate space, proper ventilation, sanitation, drainage, sufficient light should be taken care while constructing a house. Grazing area should be free from pits. Goats are kept on elevated pond dyke under widespread shady trees. Goats do not survive on

Figure 6: Goats

marshy and swampy grounds. Kids are kept under large inverted bamboo baskets until they are old enough to run along with mother.

Goats are selective feeders and relish, cumbu, napier grass (CO3, CO4), cowpea, soybean, cabbage, cauliflower leaves, lettuce, leaves of shrubs. *Acacia nilotica, A. arabica* (babul), *Azadirachta indica* (neem), *Ziziphus mauritiana* (ber), *Tamarindus indica* (tamarind), *Ficus religiosa* (papal) and mulberry, etc. are well consumed by goats.

Manure

Goat-excreta is very good organic fertilizer and contains 60 per cent organic carbon, 2.7 per cent nitrogen, 1.78 per cent phosphorus and 2.88 per cent potassium. Its solid excreta is extremely rich in nutrients than other animals. Goat-urine is also equally rich in both nitrogen and potash. Manuring with goat-droppings is advantagesous in terms of direct application in fish pond. Size of droppings is about 1.0 cm, in a shape of pellet, which is coated with mucus and floats in semi-dried conditions. The bacterial activity aggregates resulting in formation of organic detritus which is consumed by fish biomass through grazing food-chain. An adult-goat weighing about 20 kg discharges 300-400 g excreta on a daily basis. For manuring 1 ha of water area with a herd of 50-60 goats will be needed. It has been observed that Rohu (*Labeo rohita*) and Mrigal (*Cirrhinus mrigala*) grow well when pond is manured with goat excreta. This integration can produce 3.5-4 tonnes/ha/year of fish without supplementary feed or fertilizer in addition to goat-meat which has ready market throughout the country.

Chapter 8
Poultry-Fish Integration

Integrated fish-poultry farming is practiced in many countries of the world and especially in Asia. It is not only an efficient way of-recycling farm wastes but also produces high economic returns. Poultry manure is a complete fertilizer. The integration of aquaculture with poultry results in a more efficient use of resources. Other benefits of diversification include reduction in the risk of total crop failure, additional sources of food and extra income. The costs associated with fish culture operations are reduced by about 70 per cent when integrated with chicken, because fish farming recycles chicken wastes and spilled chicken feed as food and fertilizer for the fish. Consequently there is no need to provide supplemental feed or fertilizer.

Intensive production of broiler and egg laying chicken is now common in many parts of the world. The birds are typically fed with complete diets in pelleted or mash form and the manure is used fresh or as dried poultry waste.

The nutrients content of poultry manure favours its use as an agricultural fertilizer and animal feed supplement in many countries and its high value therefore warrants its usage in integrated aquaculture. Chicken raising for meat (broilers) or eggs (layers) can be integrated with fish culture to reduce costs on fertilizers and feeds in fish culture and to maximize benefits. The poultry excreta is recycled to fertilize the fish ponds and also it is used as a fish feed ingredient (dietary protein source) because it contains an undefined quantity of uneaten chicken feed crumbles. The pond embankment can be utilized for raising vegetables. Integrated farming of fish and chicken is more prevalent in Indonesia and has been adopted by certain farmers in Thailand.

This system utilizes poultry droppings of fully built-of poultry-litter for fish culture. Production levels of 4,500-5,000 kg/fish/ha could be obtained by recycling pond

Figure 7: Poultry-cum-Fish Culture

manure into fish ponds. Broiler production provides good and immediate returns to farmers. Before taking up venture it is necessary to study market demand. Success depends mainly on the efficiency of the farmer, his experience, aptitude and ability in management of flock. This involves procurement of good quality chicks, housing, brooding equipment, feeders, water-trays and management practices, and it also includes prevention and control of diseases. In India, poultry farming is mainly practiced in Andhra Pradesh, Bihar, Haryana, Kerala, Karnataka, Maharashtra, North-eastern states, Orissa, Tamil Nadu, Uttar Pradesh and West Bengal.

Poultry Farming and Management

General chicken production has shown great technological advancement and expansion during the last few decades. Poultry rearing can be divided into two general classifications.

(*a*) Layers for commercial egg production

(*b*) Broilers for commercial meat production

Total Confinement

Confinement housing is referred to a housing system in which birds have no access to any area outside the house. Generally the type of house is dependent on climatic conditions. Two distinct housing structures, based on bird management, are generally in use today. They are caged systems and deep litter or floor systems.

Caged Systems

Caged systems are used predominantly for egg laying operations but there is a growing trend towards caged broiler production. Cages contain 1 to 25 birds, depending on size. Environmental and housing conditions for this systems are given in Table 1 and Table 2.

Table 6: Poultry Housing and Environmental Requirements for Confinement Housing

Sl.No.	*Requirements*	*Layers*	*Broilers*
1.	Lighting	14 hours period	Full time
2.	Humidity	50-75% 60% optimal	50-75% 60% optimal
3.	Temperature	50-75°F	70-80° F
4.	Floor Systems:	Layer:	
	Feeder space	3-4 inches/bird	0-3 weeks : 1 inches/bird 3 weeks : 4 inches/bird
5.	Water space	0.5-1 inches/bird	0.5 inches/bird
6.	Floor space	Layer: 1.5-2sq.ft/bird	1sq.ft/bird

Table 7: For Caged Systems (Layers)

Cage Size	*Birds/Cage*	*Floor Space Inches/Bird*	*Length of Feeder and Waters Inches/Bird*
8" × 16"	2	64	4
12" × 16"	3	64	4
12" × 18"	4	53	3
24" × 18"	8	54	3
12" × 20"	4	60	3
2" × 2.5"	10	72	3
3" × 4"	20	86	2.4
3" × 4"	25	69	1.9

Litter or Floor Operations

Litter systems are used predominantly for broilers, layers and breeders. With floor operations, 4 to 6 inch of an absorbent litter material are placed on the floor prior to introduction of bird in the house. Locally available materials can be used for litter with choice. The absorbent litter materials are paddy husk, sand, saw dust, sugarcane bagasse, groundnut shells, rice bran and wheat straw. Waste management in litter system consists of periodical removal of the litter. This is done after one or several broilers have been raised, or once in a year as for egg layers. Litter is sometimes left in the house for longer periods by addition of more litter on a periodic basis. This system is known as deep littering. Litter life can be extended with greater ventilation to remove moisture.

Chicken Rearing Facility

Chicken can be raised in cages over or adjacent to the ponds. The housing for chicken are built on land rather than over the pond. The house can be made of bamboo or any other locally available cheap materials. Roof can be covered with hay or similar material. Enough cross ventilation should be maintained to keep cool during hot days. Temperature in the poultry house should always be above 20-22°C. When the temperature goes below this level, two 200-watt bulbs or two kerosene lamps for every 50 chicken can be hanged. In the case of chicken house is above the pond, the floor should be constructed with bamboo slats with 1 cm gap to allow excreta to fall into the pond but not wide enough for the chicken's feet to get caught in between and injured. The floor level should be at least 0.5 m above maximum pond water level. Many of the advantages of integration are lost when chicken are not raised over the ponds as the manure is left for long periods of time before use as they are not easily transferred to ponds, with consequent deterioration in quality.

Raising chickens over the pond has certain advantages:

It maximizes the use of space.

Saves labour in transporting manure to the ponds.

Poultry house becomes more hygienic,

No deterioration in the quality of manure as they are not left for long periods of time before use.

20 per cent of the feed (Djajadiredja *et al.*, 1980) and almost all the excreta (40g/day) (Little and Muir, 1987) enter directly into the pond water.

Each broiler bird requires one sq.ft. space and each layer bird two sq.ft. floor space area. A maximum density of 11 birds/m^2 is used for growing broilers in the Philippines (Hopkins and Cruz, 1982).

Selection of Birds

The birds may be classified on the basis of utility, economic value or fancy purpose. The kind of bird to be raised along with fish may be chosen with care from the meat type (broilers) or egg type (layers) birds. The birds of Babcock, or Leghorn breed are suitable for the purpose. Egg production and weight gains are important criteria for selection. Proper vaccination and prophylactic measures against diseases are needed for better economic returns. About 500-600 birds (layers) produce litter adequate enough to fertilize one hectare of water area. Usually eight weeks old chicks, after vaccination against viral diseases and provided other necessary prophylactic measures, are kept in poultry house near the pond.

Egg Laying

Each pen of laying birds is provided with nest boxes for laying eggs. Egg production commences at the age of 22 weeks and then gradually decline. The birds are usually kept as layers upto the age of 70 weeks. The birds lay 210 to 250 eggs/bird/year.

Feeding

Poultry feed industry is well organized sector in India and feeds are available for various stages of poultry birds under different trade names. Egg-type birds are fed with starter mash (0-8 weeks), grower mash (8-20 weeks) and brooder feed (20 weeks onwards), while broilers are fed with starter mash and 0-4 weeks with finisher feed by 4-6 weeks. Grower mash is provided to the birds during the age of 9-20 weeks at the rate of 50-70g/bird/day whereas layer mash is provided to the birds above 20 weeks at the rate of 80-120g/bird/day. The feed is provided to the birds in feeders to avoid wastage and for keeping house in proper hygienic conditions. An ample supply of water is made available at all times otherwise egg production is adversely affected.

Chicken integrated with fish culture are generally fed commercial rations which are highly balanced and has high protein content for growth and production. Fed either ad libitum or twice daily, feeds are 'diluted' by additional low cost feeds such as broken rice and maize. For broilers, starter mash is fed for 1-4 weeks and finisher mash for 5-8 weeks, and it is given as much as they can consume. A 100 kg starter mash contains 50 kg crushed maize and wheat, 14.5 kg rice bran, 16 kg groundnut oil cake/sesame oil cake, 19 kg fish meal and 0.5 kg salt. A 100 kg finisher mash has 50 kg crushed maize, wheat, 17 kg rice bran, 15 kg sesame oil cake, 16kg fish meal, 1.5 kg bone meal and 0.5 kg salt. Vitamins premix is added to both at the rate of 250-g/100 kg of feed. For chicken layers, feed is given at the rate of 80–110 g/bird/day for the first 16 weeks and 110-120 g/bird/day from 17th week onwards. Water should be provided at all times. The feed composition for layers is provided in the table.

Table 8: Feed Composition for Layers

Ingredients	*0-4 Weeks*	*4-16 Weeks*	*Above 16 Weeks*
Crushed wheat	46.0	46.0	44.0
Rice polish	20.5	24.5	29.0
Sesame oil cake	18.0	14.0	12.0
Fish meal	17.0	15.0	10.0
Bone meal	–	–	4.5
Salt	0.5	0.5	0.5

Vitamins premix at 250 g/100 kg feed.

Poultry Manure

Each bird produces approximately 40g of manure per day. Poultry manure is considered to be a complete fertilizer, combining the characteristics of both organic and inorganic fertilizers, and it has been observed that a large population of rotifers grows more quickly in chicken manured ponds than in ponds fertilized with animal manure. The conversion coefficient for chicken manure is 6.8. Chicken manure, like other organic wastes, can be converted into quality fish food, by stimulating microbial activity in the water column and at the pond bottom and realizing the nutrients and minerals originally bound in relatively indigestible form. These nutrients and minerals originally inturn provide the substrates for photosynthetic (autotrophic) and microbial production, which can be utilized by fish. The total protein content of chicken manure is as high as 10 to 20 per cent. About 80 per cent of the manure represents undigested feedstuff with 25 per cent dry matter content, which can be used directly by fish as feed. This is primarily due to the fact that chickens have a very short digestive tract, and much of their excreta is partly digested.

In India, scientists have developed a simple and economically viable system for integrating fish culture and poultry farming without the use of any supplementary feed or fertilizer. This type of farming system has been shown to produce, on an annual basis, 4500 to 5000 kg of fish, more than 70,000 eggs and about 1250 kg of poultry meat using 500 to 600 birds per ha of pond area.

Experiments and field trials in poultry and fish farming have given very encouraging results. It was observed that the excreta of 250 to 300 layer birds and 150 to 200 broiler birds produced 4 tons and 3–4 tonnes of fish per year respectively when recycled in one ha of water area. Among the carp, silver carp recorded best growth, followed by grass carp, rohu, mrigal and common carp. The per cent net return is 50.4 from fish-poultry integrated farming. The cost of fish production in this trial was much lower than that of conventional fish culture.

Table 9: Nutritional Status of Poultry Manure

Sl.No.	*Type of Poultry Manure*	*Crude Protein (1)*	*Nitrogen (2)*	*Phos-phorus (3)*	*Potash (4)*	*pH (5)*	*Crude Fat (6)*	*Crude Fibre (7)*	*Organic Carbon (8)*
1.	Fresh poultry dropping	10%	1.4–1.6%	1.5%	0.8-0.9 %	6.9	11.98%	9.59%	26% on dry weight basis
2.	Deep litter	19%	3%	2%	2%	7.8	–	–	23% on wet weight basis

In this method of farming, chicken are usually kept in confinement and hence they are susceptible to diseases and in some occasions, the whole flock may get affected, resulting in retardation of growth and reduction in egg production. In some cases, broilers do not attain market weight of 1.5-2 kg in time and sexual maturity is delayed in layers. Therefore, protective measures in the form of preventive vaccination (for Raniket or Newcastle disease, fowl pox and infectious buccal disease) and curative treatment (for parasitic diseases) should be undertaken.

Harvesting

Eggs are collected twice a day, *i.e.* morning and evening. Layer birds are discarded after 18 months of rearing as eggs production goes down. Further rearing may not be economical. Marketing of broilers starts after 5-6 weeks of rearing, during which birds weigh 1.4-1.6 kg.

Limitations

Fish ponds fertilized with poultry manure are subjected to variation in their loading, as food intake and waste output increases over the life cycle of the birds. This variation is greatest with meat or broiler bird production. These fluctuations in waste input can be reduced by using a production flock of mixed age. Most farmers may find it difficult to sustain the large number of birds due to non-availability of locally available feed ingredients for poultry. Limited levels of integration may be achieved by rearing few birds and letting them roam freely avoiding the need for purchasing concentrate feed. Availability of vaccines and correct curative treatment for chicken diseases is a liability factor in meat or egg production. Finally, requirement of investment and managerial skills may limit the acceptance of chicken-fish culture by farmers.

Advantages

Raising chicken over ponds has a number of benefits, and are as follows:

1. Chicken houses constructed on ponds do not have to compete for separate price of intended land for other purposes,
2. Hygienic conditions are better in chicken houses constructed over ponds, as the droppings fall directly in to the ponds.
3. Chicken excreta provide food and fertilizers for fish culture.

Chapter 9
Duck-Fish Integration

The first scientific experiments on duck-fish farming were made by Probst in Germany in 1934, but because of World War II the results remained unutilized. After the war, when there was a serious protein shortage in European countries, large-scale experiments were initiated in Hungary (1952), Czechoslovsakia (1953) and East Germany (1955) to determine optical husbandry methods for raising ducks on fish ponds in the climatic condition of central Europe. This method of integration spread to other Asian countries like India, Thailand, China, Bangladesh, etc.

The combined culture of ducks and fishes can be justified that, they are mutually beneficial and the profitability of the integrated system in turn increases significantly. The foremost benefit of the combined culture of fish and ducks is that considerably more animal protein can be raised on the same area. Moreover, the ducks benefit from the pond and fish growth is accelerated by the duck manure, which provides a continuous supply of organic matter containing carbon (C), nitrogen (N) and phosphorus (P). These organic matter boost biological production in the water column and increase the natural food supply for the stocked fish. The ducks are living 'carbon-manuring machines'. Duck-fish integration is the most common type of integration in China, Hungary, Germany, Poland, Russia and to some extent in India.

The manuring effect of ducks is highest when their resting and feeding places are fixed over the pond surface, but this brings higher labour requirements for giving feed. Therefore resting and feeding places are often provided on flat areas of the pond shore. If the ducks are of a suitable strain and are habituated to roam over the whole pond surface, approximately 50 to 60 per cent of their manure will fall into the pond. They seek natural food in the pond and consume wide variety of organisms, *e.g.*, tadpoles, frogs, insect, insect larvae, snails and waterweeds. The ponds provides ducks with a healthy environment.

The rationale for raising ducks on fish ponds is stronger than many other animal/fish systems, since the pond can provide living and foraging area for the ducks as well as the fish. Duck eggs are an important source of food in India. Consumption as well as production of duck eggs in India is mostly done by socially weaker sections of the community. Though Asia is considered to be the land of domesticated ducks, the best breeds and strains available have been developed for their excellent egg/meat production in Europe and America through systematic breeding, feed management and disease control. This integrated farming system of combining ducks with fish has considerable scope in West Bengal, Assam, Kerala, Tamil Nadu, Andhra Pradesh, Bihar, Orissa, Tripura and Karnataka.

Methods of Raising Ducks

The following methods of raising ducks are adopted.

(*a*) Extensive Raising

In this system, the ducks are left out to graze and only small amounts of supplementary duck feed is provided and the number of ducks is limited by the food they can find in the pond water. The amount of manure contributed to the pond and its effect on fish yield are also limited. This method is usually employed in Europe, where approximately 150-500 ducks are reared per hectare of pond surface.

(*b*) Intensive Raising

The ducks are fed with concentrates at the same rates as reared on land. The stocking density of ducks per unit of pond area is also high. Therefore, higher amounts of manure and uneaten duck feed (estimated to be 10 per cent) are loaded into the fishpond, and consequently higher yields can be obtained (Hepher and Pruginin, 1981). This method is usually employed in Africa, where duck stocking densities are approximately 1000 – 2500/ha.

(*c*) Semi-intensive Raising

While extensive and intensive methods of duck production rely completely on natural feeds and concentrate feeds respectively, with no reliance on fertilization of ponds, the semi-intensive mode is characterized by fertilization of ponds to produce natural feeds (Edwards, 1985). Ducks graze in fertilized rice fields or fishponds in addition to wild sources of natural feeds.

Types of Integrating Practices

The integrated farming system is pursued by adopting two types of practices *i.e.*, open and closed system (Singh *et al.*, 2002).

(*a*) Open System

In open system of integration, the ducks are allowed to go into the pond for a certain period of time every day. It is found suitable for small individual ponds.

(*b*) Closed System

The closed system of integration is most complex in comparison to open system of farming. The duck house or cages are built on the pond embankment or over the pond margins so that droppings can go directly into the pond from the floor of the cages. By adopting this system, energy losses as a result of transportation is avoided (NACA, 1989). It is very much successful in well-managed large farms.

Housing and Management

The ducks normally do not need elaborate housing system. Since the ducks remain most of the time in the pond. A low-cost house with adequate accommodation, reasonably cool in summer and sufficiently warm during winter, with sufficient sunlight and protection from rain should be made available for ducks. It should give protection to birds from their natural enemies like jackals, fox, dog, cat, snake, crows, etc. The ducks may be reared in an open-water or in a centralized system or in a floating house or house constructed on pond dykes. In open-water, the ducks are left in large numbers for grazing in lakes and reservoirs during the day-time, while during night time they are confined in duck house. In centralized system, relatively large duck sheds are constructed in vicinity of the fish-ponds with cemented area of dry and wet run. The average stocking rate is about 4 ducks/m^2. Dry runs are cleaned on daily basis, while wet runs are cleaned at an interval of 2 or 3 days and fertilized water is flushed into fish ponds. This method is of advantage for its centralized management mechanism. The left-over feed is not fully utilized in this system and is also unable to take advantage of duck-fish symbiosis. Yet another method which is very common is construction of duck house on pond dykes. Floating duck houses are also very common. The dykes of grow-out ponds are partly fenced to form a dry run and part of water-body is fenced to form a wet run. Generally net pen is installed 40-50 cm above and below water surface. Ducks can enter wet run for food but cannot eat fish. A cheap house is generally used for rearing ducks. Floating ducks house can also be made using empty oil-barrels.

Jhingran and Sharma (1980) reported that in India, the ducks were allowed to range freely over the pond water during day time and sheltered at night in a floating duck house. The floating duck house is made of bamboo matting over empty oil drums, positioned close to the pond bank, where the young ducklings and breeding birds were usually reared. In Thailand, duck houses have been constructed over the pond using only bamboo; a fence of bamboo strips enclosing approximately half of the pond area confined the ducks over the water, but allowed fish access to the entire pond. Edwards (1983) recommended five ducks per m^2 of floor space with this type of housing. Nylon netting has also been used for fencing but this was found to be less durable and effective. Regular maintenance of duck fences and houses is required. Ducks should be restricted to a

Figure 8: Duck-cum-Fish Culture

relatively small area of pond if they are being raised in the intensive mode, to avoid the expenditure of too much energy in swimming (Edwards, 1985).

Breeding Stock

Different duck strains are used according to traditional practices and availability (Little and Muir, 1987) in various countries.

1. In India Khaki – Campbell – Bengal runner cross is favoured. (Jhingran and Sharma, 1980).
2. In Taiwan, the ducks raised are either the egg producing native duck or the meat producing mule duck (a cross of the native mallard duck and the drake of the muscovy (Chen and Li, 1980).
3. In Thailand, the Khaki-Campbell (Anas platrhchos) is raised (Lawson, 1981).

4. Sin (1980) reported that in Hong Kong either the local introduced Thai breed and/or a Taiwanese hybrid (Male peking-female Denmark cross) is raised.
5. In the Philippines and in the Central Europe, the Peking duck and the various hybrids of this strain are the most commonly cultured (Little and Muir, 1987).

Ducks may be raised specifically for meat and egg production depending upon the demand of local market. Edwards (1983) considered egg laying Khaki-Campbell type ducks as being most suitable for small-scale subsistence production in Thailand on the basis that the ducks will furnish a daily supply of good quality egg protein (*i.e.*, eggs). The surplus eggs are easier to sell than whole ducks, and ducklings need to be replaced only every 18 months, as opposed to three months for fast maturing meat ducks. In addition, an even manure input is maintained over a longer period, which is beneficial for fish production.

The improved breed of 'Indian runners' has been found more suitable in view of its hardness. It has been found that 450-500 ducks are sufficient to produce manure required to fertilize one hectare of water area. The ducks are fed on superior quality of poultry-feed and rice-bran in 1:2 ratio by weight at 100 g feed/bird/day.

Management of Ducks

Day old ducklings and 10-14 days old ducklings are ideal for stocking. The minimum number of breeders should be about 1000 to 2000 ducks of 1 to 2 year of age, with sex ratio of 1 male to 4–5 females. The breeders lay eggs up to 4 years of age but with decreasing returns. Sexual maturity is achieved at 5 months and egg laying commences at 6 to 7 months. A suitable protein rich feed (20 per cent protein), clean drinking water and proper illumination are mandatory. The daily feed ration is about 9 to 10 per cent of the body weight. About 120 to 140 eggs per female duck are produced. The normal egg weighs about 80 to 90 g and is 8cm long and 5cm wide. The incubation period is 28 days. The survival rate of day old ducklings is 75 per cent. Losses during the nursing period (10 to 14 days) and rearing periods are 8 per cent and 2–3 per cent. Day old ducklings require special care and a controlled environment with respect to temperature, feed, drinking water, light, space, etc., upto an age of 10 to 14 days, after which they can be stocked on the ponds. During the first week, about 50 to 55 ducklings per square metre are kept in a heated room on a mesh floor (0.2cm thick wire; 1.5cm mesh size) to allow manure to fall through into the pond. The pelleted starter feed is offered through feeders and clean drinking water is provided in troughs, which permit access only to the beak to prevent ducklings from becoming wet. The air temperature must be about 30 to 32° C.

Keeping the Ducks on Suitable Ponds

In commercial duck fish farming practice, the ducks were allowed to roam freely on large ponds, distributing their manure uniformly over the whole pond area while foraging for natural food. Wire enclosures are provided for the ducks in fish pond (called "water runs") with suitable resting and feeding places on the shore or over the water. The enclosures keep the ducks in, but allow the fishes to enter the "water runs". Wave action carries the well manured water out of the "water runs". About 0.5 to 1.0m^2 of enclosed dry run and 2 to 4m^2 of 'water run' is allowed per duck.

Feeding

Feeding is an important component, as first three weeks of ducklings are vital for future growth. Feed should have sufficient protein. The starter ration should have vitamins, minerals and trace elements. Feed with 17 per cent of protein and high level of energy will be excellent for proper growth of ducklings, while a ration with 15 per cent protein will be good enough for layers. Ducks are voracious feeders and in addition to concentrate feed it also consumes snails, fingerlings, earthworms, aquatic insects and aquatic weeds. Ducks find it difficult to swallow dry mash. Pelleted feed is better which reduces wastage. Several authors have found that ducks will consume frogs, tadpoles, mosquito and dragon fly larvae and aquatic weeds which are generally not consumed by commonly stocked fish. Diets containing a large proportion of paddy rice were found to be unsuitable for high yielding Khaki-campbell type ducks which had access to little extra food from the ponds. A diet better balanced with respect to protein, energy and fibre has to be formulated. Feeding twice a day, (0.14 kg/duck/day) a stable egg laying rate of 80 per cent was achieved under small farm conditions.

Duck Feed

The ducks are fed using demand feeders. The general composition of feeds for the different stages of duck rearing are given in Table 10.

Table 10: Composition of Feeds for Ducks

Components (Values in %)	*Type of Feed*			
	Starter	*Rearing*	*Fattening*	*Layers*
Dry matter	86	86	86	86
Carbohydrates	68	68	69	67
Digestible protein	18.0	15.0	11.50	16-8
Crude protein	20.50	18.0	14.0	19-20
Calcium carbonate	3	3	3	4.5
Calcium phosphate	0.8	0.7	0.7	1.6
Vitamins premix	0.5	0.5	0.5	0.5
Minerals premix	0.5	0.5	0.5	0.5

Table 11: Requirements of Ingredients for Duck Feeds

Ingredients	*Starter Feed (%)*	*Rearing Feed*	*Layer Feed (%)*
Maize	20	20	20
Wheat	57.9	53.9	40.9
Fish meal (64% crude protein)	10	2	2.5
Soya grit (47% crude protein)	8	9.8	12.3
Meat meal (50% crude protein)	0.9	4	5.7
Wheat bran	2.5	8.7	5.0
Food lime	–	0.3	5.3
Methionine premix	–	0.5	0.5
(5 per cent active ingredient) salt	0.2	0.3	0.2
Vitamins premix	0.5	0.5	0.5
Minerals Premix	–	–	2.5
Wheat germ	–	–	1.5
Milled sunflower seed cake	–	–	3.1

Diseases

Ducks are susceptible to aflatoxins and hence feed ingredients should be stored properly to prevent aflatoxin contamination. In addition to this, ducks plague, viral hepatitis and cholera may cause heavy losses in duck population. Timely vaccination is of tremendous help against viral diseases. Restlessness and water discharge from eyes and nose is symptom of disease. Isolation of bird is the first step to prevent spread of diseases.

Egg-Laying of Ducks

The ducks start laying eggs after attaining 6 months. Khaki Campbell breed is a prolific layer and starts laying eggs at the age of 3 months itself and the process continues till 2-3 years, depending on the duck species, nutrition, health and environment. Since laying of eggs is done at night, there is no possibility of an egg being laid when birds are in ponds during day-time. It is advisable to keep some

Figure 9: Ducks in the Pond

straws or hay in corner of duck house to facilitate egg-laying. The eggs are collected every morning. Considerable attention needs to be paid to pen hygiene.

Duck Manure: Chemical Composition and Fertilizer Potential

Duck manure had been reported to contain approximately 57 per cent water and 26 per cent organic matter. Each duck produces about 7 kg fresh manure over a 36 day period and 500 ducks therefore produce about 3.0 to 3.5 tonnes in the same period. The manure contains 57 per cent water and 26 per cent organic matter and 100kg contains about 10kg carbon, 1.4kg P_2O_5 1kg N, 0.6 kg potash (K_2O), 108kg calcium and 2.8kg of other materials. The effect of stocking ducks on fish ponds, or the regulated use of their waste, will vary widely with their age, size and production method. Fast maturing meat ducks will have an increasing manure output during their growing period; this should be managed to avoid over addition to the pond and consequent problems of fish production. The management of a mixed age flock of ducks can alleviate this problem. Conversely, egg laying ducks will supply a more constant input of waste into the pond since their feed input and manure output is more constant.

Overloading of duck manure in the ponds should be avoided during winter season. Fish mortality may occur occasionally due to a continued build up of waste during the winter season, causing a subsequent bacterial and plankton bloom because of temperature rise, finally resulting in depletion of dissolved oxygen from water (Singh, 1980).

The duck manure has a pronounced effect on natural feed production *i.e.*, plankton production in contrast to ruminant manure, which inhibits phytoplankton growth. The application of duck manure and grazing by ducks in the ponds improve the physico-chemical parameters of soil like pH, organic carbon percentage, phosphorus and potassium. In addition, parameters like alkalinity, dissolved oxygen, free carbon dioxide, conductivity, $KMnO_4$ consumption and plankton concentration also improves. The change in different parameters of soil and water indicate that ponds become more productive. The detritus formed by duck excreta at the pond bottom serves as a substrate for micro-organisms as well as food for zooplankton and fish. The plankton level in ponds manured with duck droppings increases. The microbial community in detritus is known to provide essential nutritional requirements to the fish feeding on it.

Harvesting

The fish-rearing period is generally kept as one year and yield ranging from 3,000 to 4,000/kg/ha/year is expected and in addition to this, eggs and duck-meat are also obtained in good quantity on an annual basis.

Chapter 10

Paddy-cum-Fish System

Paddy is the major crop and rice is the staple food for over 1.6 billion people of the world. Over 90 per cent of the paddy is produced in Asian countries, and it is the sole livelihood of most of the rural farmers. Collection of wild and naturally occurring fish from paddy fields has been an age-old practice as the rice cultivation by itself. In fact, the practice of incorporation of fish in rice fields was introduced in South East Asian countries from India about 1,500 years ago. Planned paddy-fish system ensures higher productivity, farm income and employment in these areas. Paddy-fish integration provides a net annual income which accounts for several fold increase over traditional practice. It also facilitates crop diversification, thereby reducing investment risk and also generating year-round employment opportunities in farm.

The paddy fields retain water for about 3-8 months in a year, and paddy-cum-fish culture can provide an additional supply of fish crop. The culture of fish in paddy fields, which remain flooded even after the paddy is harvested, might also serve as an off-season occupation for farmers. Paddy-cum-fish culture needs modification of rice-fish plot, digging of peripheral trenches, construction of dykes, pond refuge, sowing improved varieties of rice manuring, stocking of fish at 10,000/ha and finally feeding of stocked fish with rice-bran and oilcakes at 2-3 per cent of body weight. Harvesting is done when fish attains marketable size. Vegetables, fruits-bearing plants, papaya and banana, can be grown on fallow land for an additional production. Such system provides a net income of about Rs.35,000-40,000/ha/year, which is much higher over traditional practice. Growing fish on the same piece of land as paddy has several distinct advantages as follows:

Intensifies land and water use.

Contributes to increased and diversified food supply.

Increases returns on the use of land.

Serves as an alternative in areas where fish culture in ponds is not feasible for reasons such as lack of land tenure or high capital cost and risk.

Stocking of fish in paddy fields effectively uses the available food and space.

Controls weeds when herbivorous fish is stocked within one week of transplanting paddy

When stocked at the start of paddy production, fish feeds on competing phytoplankton.

The activity of feeding fish and their excretion improves paddy fertility.

Minimizes additional labour requirement.

If crop is long maturing, simultaneous fish culture provide an earlier financial return.

Efficient production of high carbohydrate (rice) and high protein (fish) food from same area of land.

Fish acts as a biological control of paddy pests namely stem borers, plant and leaf hoppers.

Higher level of water management for fish improves conditions for plant crop.

Reduces production costs as paddy bottom becomes soft and clean after fish culture.

The rearing of fish along with paddy is carried out by two methods:

(*a*) Capture Method

No stocking of fish takes place. It consists entirely of indigenous fish populations entering flooded ponds and paddy fields, where reproduction takes place. Harvest is done on fish entering, during and at the end of crop grouping season. This method is widely adopted when compared to the other methods. It is prevalent in Hong Kong, India, Indonesia, Japan, Malaysia, Thailand and Vietnam.

(*b*) Culture Method

Stocks of fish are deliberately introduced. Fish culture in paddy fields may be concurrent with rice culture or in rotation. In concurrent type, fish is reared simultaneously with the growing of the plant crop. The fish increases the plant production by fish residues and it also helps to control pests, weeds, insects and mollusks. In rotational culture, fish may be cropped either after a single annual rice crop or between harvest and next replanting. Modern intensified cropping method using pesticides and fertilizers is used. It reduces water needs and pesticide accumulation in fish. It controls the pest by interrupting their life cycles. The plant residues (*e.g.* rice stubble, split grain) benefit the fish. Also, the fish residues benefit crop production. The production costs also reduces since paddy bottom becomes soft and clean after fish culture.

Features of Paddy Fields

10-20 cm water depths depending on the size of the fry or fingerlings of fish

Use of a trench system (30-40 cm deep and 50-70 cm wide)

60 cm dams for isolating the rice fields

Water inflow and outflow with different size mesh nets.

Rice-fish culture system offers an environment in which the rice farmer can produce additional food or saleable product. Knowledge of the seasonal presence of food organisms for fish is important so that the farmer can asses the field quality, determine optimal stocking densities, sizes and species and whether supplemental feeding is necessary.

It is accepted that the stocking of fish in paddy fields more effectively uses the available food and space. This is particularly true if a polyculture of fish is used. The stocking of fish to feed on competing phytoplankton and submerged weeds benefits the rice, particularly if done at the start of the rice production. The stocking of macrophagous tilapia within a week of transplanting rice was claimed to control weeds by Mortimer (Pullin, 1983).

Researchers in China suggest that this fertilization effect is more marked in less fertile fields and can be cumulative over a few years (Spiller, 1985). Hora and Pillay (1962) reported that the yield of rice increased approximately 15 per cent on an average in the Indo-Pacific countries when fish are introduced. Khoo and Tan (1980) reported that in China and Russia, increase in rice yields of upto 14 per cent and 7 per cent have been recorded. By adopting polyculture method in rice fields, fishes such as *Orechromis niloticus, O.mossabicus, Cyprinus carpio, Puntius goniotus, Helostoma temmincki* and *Osteochilus hasseltii* can be cultured at different ratios.

The time table of rice/fish culture is variable depending on the type of fish required and method (concurrent/ rotational). In general, stocking of fry occurs around ten days after transplantation of rice from the nursery to the fields and fingerlings after three weeks (Singh *et al.*, 1980). The rotational culture of rice and prawns in paddy fields that are seasonally brackish has been reported in Bangladesh.

Paddy cum fish culture is an age old practice in India and other South East Asian countries. About 2.3 hectare of low-lying paddy growing tracts occur in the Eastern and Northeastern states of India covering Assam, Manipur, Tripura, West Bengal, Orissa, Northern Bihar and Eastern Uttar Pradesh where one crop of tall deepwater variety of rice is only cultivated.

For a renovated paddy plot of 1.02 ha, a perimeter canal of 0.27 ha is excavated inside the plot. Fingerlings of Indian major carp, rohu, catla and mrigal at the rate of 6,000 nos. ha in the ratio of 3:3:4 in the same field. The fishes are fed with rice bran and groundnut oil cake at 1:1 ratio at the rate of 2-3 per cent of body weight. After harvesting the deep water paddy, the fishes will move into perimeter canal instinctively. During the next season, high yielding varieties of rice are cultivated in the same field. Before applying pesticides, care should be taken to prevent drainage of pesticide washings into the perimeter canal by making a long dyke along the periphery of the paddy plot. The deep water paddy variety yields 1200 kg/ha whereas the high yielding variety (Ratna and Jaya) rices yield 4,300 kg/ha making a total of 5,500 kg/ha/yr within 10 months and a production of 700 kg of fishes/ha could be achieved.

The integration of paddy with fish in bilaterally renovated paddy plots can be done through recycling of domestic wastewater. For this, carp culture with paddy is taken up in 0.38ha paddy plot having two ponds (0.07 ha each) on either side or a central portion (0.24 ha) for paddy cultivation.

Fish (catla, mrigal,rohu and common carp) can be cultured at a stocking density of 5,000 nos/ha and fish

Figure 10: Paddy-cum-Fish Culture

production of 956 kg/ha within 9½ months. Total paddy (deep water and high yielding variety) production was 6,814 kg/ha.

The trapping of shrimp larvae in fallow paddy fields and growing to market size is an age-old practice still existing in several parts of India.

Rice-*Azolla*-Fish Integration

The rice-fish-*azolla* mixed culture were initiated in Philippines in the mid 1980's. Initial focus was given to assess the effectiveness of fresh *azolla* supplement for giant tilapia (*Oreochromis nilotica*). Researches were concentrated in developing simple processing techniques for converting excess azolla biomass in a farm that could be stored and used as a major ingredient of fish rations, a supplement comparable to commercial feeds in quality. Fish species grown concurrently with low land rice, feeds heavily on fresh or processed *azolla*.

When about 30 per cent of the inorganic nitrogen fertilizer for rice was substituted with azolla, it resulted in slight increases in net income. The feed formulations using *azolla* meal (finely ground dry azolla) as the major component was found to be as effective as commercial feeds. Fresh *azolla* given in unlimited quantities in addition to known amounts of rice bran was less effective. Integrating rice-*azolla*-fish with low land rice produced significantly higher income than rice monoculture (Jovena and Labitan, 1997).

Cultural systems are often very important sources of fish where traditional rice growing methods are used. Naturally occurring fish of economic value found in paddy fields tend to be slower growing and/or predatory. With adaptation, they have potential for some intensification and subsequently higher yields and minimum capital inputs. For this, the rice farmer's attitudes needs to change from that of the hunter to fish farmers.

Cultural systems for raising fish-in-rice benefit the rice crop and provide a better production of fish than captural methods. In the management of fish-in-paddy culture system, every effort is made to remove the wild fish and prevent them from entering the culture area, to avoid predation on the more valued stocked fish. However, these fishes are preferred for reasons of taste and value. Cultural methods vary according to the climate, water availability, fish species, plant variety and traditional custom.

Production of fry, finger lings and market-sized fish are all practiced in rice fields. Fry and fingerlings require a much shorter culture time and are more suitable when rice production methods require regular draining of the paddy. Also, where rice production is almost year round, a crop of fry or fingerlings are quickly taken in between. Climate and water availability may make the production of food fish difficult, depending on the accepted marketable size. Supplementary foods are given in more intensive systems, especially when the culture period is limiting. Such feeding allows faster individual growth to ensure that the required sized fish are produced. However, high levels of feeding can also be counterproductive as it may be wasteful and/ or cause deterioration in water quality.

Chapter 11
Plant-Fish Integrated Systems

Plant materials used as fish pond inputs can be of terrestrial or aquatic origin. They may be grown specifically for use as fish feeds or used as and when they become available or gathered from the wild. Sometimes plant residues are used as pond inputs, the major crop being reserved from direct human or animal feed. There are limitations to the inclusion of plant materials as complete feeds for intensively raised fish but they are very useful as supplementary feeds in more extensive systems. Farmers use various areas of their farms, including fields; small plots of unused land, pond dykes and drained ponds. Available water resources such as rivers, lakes, ditches and pools are also used to cultivate aquatic plants. Plant materials are:

Used as a direct feed for herbivorous macrophyte feeders in pond fish culture.

Added to ponds without stocking fish, vegetation decomposes and acts as a fertilizers or 'green manures'.

Used as organic fertilizers during the preparation of the pond prior to filling.

The remainder of both coarse and soft materials of crops of grains and legumes where the 'tops' are fed to cattle or used as feed for fish.

The feeding of plant materials is particularly relevant when animal manures are not available in sufficient quantities, either through a shortage of livestock or through competition with other users.

The use of plant materials in fish culture has considerable potential where field cropping generates and excess of crop residues and by-products, which have a low economic value and high bulk, and whose alternative uses are hard to find.

Manured pond water can be used for irrigation.

Grass species, easily produced on the farm serve as low-cost supplemental feeds for fish.

Used to stabilize pond embankments, and

An additional source of income.

Because of the different varieties of crops, fish-plant integration is subdivided into three main models:

(*a*) Fish-Grass

Pasture grasses are grown in unoccupied fish ponds and the draw-down area of lakes and reservoirs. The grasses are cut to feed fish. Later when the water rises, the remaining grasses are submerged and serve to fertilize the production of aquatic organisms. These water are fertile for either fingerlings or grow-out fish.

(*b*) Fish-Terrestrial Plant

This is the most common model. Pond dikes and odd pieces of land are fully utilized to grow plants as fish feed. Sometimes, fruit trees are interplanted in crop fields.

(*c*) Fish-Aquatic Plants

The rivers, lakes, ditches, ponds and low lands are partially used to grow aquatic plants as fish feed.

Grass

Grasses are cultivated on the wide dykes between ponds exclusively as fodder for herbivorous fish. Elephant or Napier grass (*Pennisetum spp.*), English rye (*Lolium perenne*) and Sudan grasses (*Sorghum sudanenese*) grown in the river areas are cut repeatedly, fertilized and irrigated with pond sediments and water (Edwards, 1982a). Other grasses for fodder includes guinea grass (*Panicum maximum*) in Thailand and Ialang grass (*Imperata rundinacea*) in Malaysia (Little and Muir, 1987). Star grass (*Cynodon plectostachyus*), which is productive under dry conditions, an advantage for fish culture in regions of seasonal rainfall, are grown in Africa and Asia. The barnyard grass (*Echinochiao crusgalli*) is cultured in dry nursery ponds prior to their use for raising Chiese carp fry.

Plants

A variety of vegetables are grown for feeding, particularly the cabbage family (*Brassica* spp.), maize (*Zea mays*), sorghum (*Sorghum spp.*) and sweet potato (*Ipomoea batatas*) for feeding fish. A mixture of perennial and annual crops is also grown. It includes a variety of vegetables (sweet onion, green onion, sweet potato, water cress, tomato, cabbage, water spinach, papayas, pumpkins, gourds, spinach, brinjals, cucumbers and leafy vegetables), fruits (banana, orange, peach, litchi, longan, apple, apricot) and other crops (sugarcane, tea and cassava). Pond mud is annually removed and used as manure in fruit trees.

The plant crop residues or manioc (*Manihot esculenta*), sweet potato (Ipomoea batatas) leaves, young leaves from sugarcane (*Saccharum officinarum*), leaves of Alocasia macroorhi, vines of the velvet bean (*Mucana spp.*), leaves of papaya (*Carica papaya*) and banana leaves (*Musa spp.*) are used as herbivorous fish feeds in various parts of the world.

Where embankments are wide enough, mulberry and bamboo are grown. The mulberry leaves are used as feed for silkworms. The silkworm excreta and silkworm sloughs (moulted skins) both serve as feed and fertilizer.

The produce from bamboo farming mainly bamboo shoots. Wastes and by-products from bamboo shoots are used for fish farming. They can be used as firewood, construction materials for livestock pens or support materials for climbing plants. For shoot production, compound fertilizers containing N,P,K are needed which are provided by the pond mud.

Aquatic Macrophytes

Aquatic macrophytes, as distinct from phytoplankton, are considered as those higher plants which grow in water continuously. They may include those present in soils covered with water during a major part of the growing season. The abundance of aquatic macrophytes in many parts of the world has created a dilemma. Often considered as a weed problem to be controlled, possibly using fish, aquatic macrophytes can in suitable circumstances be viewed as a highly productive crop requiring no tillage, seed or fertilization (Ruskin and Shipley, 1976). Edwards (1980) has reviewed the pathways in which aquatic macrophytes may be used in food production, and has detailed potential integration with fish culture. Fish show preference for different types of aquatic macrophytes based on succulence and taste, as with plant materials in general, and this will affect their usefulness in weed control. Least favoured by grass carp, for instance are the rushes, sedges, watercress, water lettuce and water hyacinth. Most favoured include the filamentous algae, soft submerged macrophytes and duckweeds. Aquatic macrophytes are used as feed for fishes. Other potential uses broadly include their use as fodder for livestock, recycling wastes, as fertilizers and as direct human food.

The Chinese cultivate useful water plants such as *Eichhornia crassipes* (Water hyacinth), *Pistia stratiotes* (water lettuce) and *Alternanathera phylloxeroides* (alligator weed) in rivers and canals adjacent to fish farms, and they form a major source of feed for both livestock and fish (Edwards, 1982). Duckweeds are grown in small ponds and are harvested by skimming or seining with a net and fed to herbivorous fish (Little and Muir, 1987). In Indonesia, *Lemna minor, Hydrilla verticillata* and *Chara spp*. are grown for feeding Nile tilapia (Rifai 1979, 1980).

The residues of aquatic macrophytes grown principally for human consumption are also used as pond inputs. The growing tip, with fresh young leaves and stem of water spinach (*Ipomoea aquatica*) is cut for human consumption and the remainder is used as a fish feed.

Considerable requirements for space (adequate land area for growing large amounts of grass to adequately feed the fish) and labour (cutting grass for the fish pond, removing mud and broadcasting on to grass) may be a limiting factor. The resource costs of growing the necessary high quality pasture grasses needs to be considered.

Use of Aquatic Macrophytes in Pond Systems

(*a*) Culture of Food Plants and Fish in the Same Pond

The culture of either submerged or floating aquatic macrophytes for feed with herbivorous fish together in the same pond is not normally an effective approach. Fertilization of the pond to encourage the necessary growth of submerged macrophytes, would rather tend to favour phytoplankton growth. In turn, the submerged macrophytes would be eliminated by their shading effect (Edwards, 1980).

Usually, floating macrophytes cannot be maintained in balance with a population of growing herbivorous fish. Conversely, if a lower stocking density of fish is used, problems may arise with the macrophytes covering the pond surface excessively.

(*b*) Cultivated Aquatic Macrophytes

The Chinese cultivate useful water plants in rivers and canals adjacent to fish farms, and they may form a major source of feed for both livestock and fish. Three aquatic plants in particular are grown. *Eichhornia crassipes* (water hyacinth), *Pistia stratiotes* (water lettuce) and *Alternanthera phylloxeroides* (alligator weed)

Duckweeds may also be conveniently grown in small ponds, harvested by skimming or seining with a net, and fed to herbivorous fish. Duckweeds are grown in shallow ponds converted from rice paddies on the outskirts of some urban communities in Taiwan. The duck-weed is harvested weekly; ponds are skimmed with bamboo poles and the concentrated duckweed sieved into sacks and sold fresh, for carp fry feed. Importantly, only a proportion of the duckweed must be maintained, otherwise the pond becomes dominated by phytoplankton and the duckweed diminish (Chao, 1983). Van Dyke and Sutton (1977) reported efficient digestion of *Lemna* spp. fed to grass carp, and it appears to be a favoured food. Apart from high productivity, duckweeds have favourable protein, fat and fibre contents, particularly when cultured in nutrient rich waters. In Indonesia, *Lemna minor* was found more suitable than the coarser *Hydrilla verticillata* and *Chara spp*. for feeding to the Nile tilapia in cages (Rifai, 1979, 1980).

The use of *Azolla pinnata*, the aquatic fern that has a symbiotic relationship with a blue-green alga (*Anabaena azolla*), as a fish feed has been reported. The fern has been cultivated in rice fields as a green manure, and also used to supplement the diets of pigs and poultry. Its rapid growth (the fern doubles its biomass in 3-10 days) and high protein content suggest a use for it in integrated fish culture

Implications of Integrated Agriculture-Aquaculture System

Integrated vegetable production and fish farming could both improve household nutrition and increase cash incomes dramatically. To date, the analysis has been confined to carbohydrates and proteins. Proximate minerals and amino acids are expected to demonstrate that the integration of vegetable fields and ponds perform an even more important nutritional function.

It is important to note that vegetable production demonstrates much stronger nutritional and economic benefits than fish farming. Likewise, fish culture ponds function as a mini reservoir for irrigation purpose during dry seasons. On the other hand, its organically enriched pond mud is a good fertilizer for crops and can be used as a sediment trap to catch friable soils eroded from upstream during wet season. It could also act as a buffer to reduce wet season rainfall run-off and could reduce the flooding.

Farm ponds play a variety of roles simultaneously in the integrated management of small farm resources and they are not merely fishponds. However, in integrated farm resource management, fish in farm ponds function as and essential catalyst in the two principal symbiotic ecological pathways between the pond and the vegetable field *i.e.*,

1. Vegetable waste enriching the pond water and
2. Pond mud and nutrient enriched water for irrigation.

Horticulture–Fish System

Fruits and vegetables are rich in nutritive values and contain carbohydrates, fats, proteins, vitamins and minerals. Floriculture trade is well established in the USA and European countries. In India too, it is a booming industry for loose and cut-flowers. Horticulture as well as its related industries are excellent avenues for employment generation and as a whole for earning foreign exchange. These crops can be increased by bringing more area under

Figure 11: Banana-cum-Fish Culture

cultivation. Ponds are well suited for the purpose. In India pond-dykes are usually not used, but in China these are being used for multipurpose production. The top, inner and outer dykes of ponds as well as adjoining areas can be best utilized for horticulture crops. Pond water is used for irrigation and silt, which is a high-quality manure and contains several nutritive elements, as base manure for crops, vegetables and fruit bearing plants.

The success of the system depends on the selection of plants. They should be of dwarf type, less shady, evergreen, seasonal and highly remunerative. Dwarf variety of fruit bearing plants like mango, banana, papaya, coconut and lime, are suitable, as these plants do not obstruct sunlight to water-bodies. And pineapple, ginger, turmeric, chilly are grown as intercrops. During summer, brinjal, tomato, chilly, gourds, cucumber, water-melons and ladies' finger, while during winter peas, cabbage, cauliflower, carrot, beet, radish, turnip and spinach are grown. Plantation of flower-bearing ornamental plants like tuberose, rose, jasmine, gladiolus, marigold and chrysanthemum provides additional income to farmers. Farming practices are carried out on broad dykes, which can stand ploughing and irrigation. Ideal management involves utilization of middle portion of the dyke, covering 2/3rd of the total area and rest of the area along the length of the periphery for intercropping of vegetables. Residues of vegetables cultivated could be recycled into fish-ponds, particularly when stocked with fishes like grass carp. Monoculture of grass carps at stocking density of 1,000/ha has been observed to give a production of about 2,000 kg/ha/year and addition of common carp in this practice has been found to be beneficial for utilizing resulting faecal debris. In mixed culture of grass carp along with rohu, catla and mrigal, in 50:15:20:15 ratio at a density of 5,000 fish/ha yielded fish to the tune of 3,000 kg/ha/year. This integrated system fetches about 20-25 per cent higher returns compared to aquaculture alone, beside generating employment opportunities round the year.

Mushroom–Fish System

Mushrooms are fleshy fungi and are one of the most favoured food items. Consumption of mushrooms as food and medicine is recorded even in the classical religious writing, Vedas and Bible. More than 1,000 species of edible mushroom are reported and about 200 species are on record. They are rich source in India of protein as compared to cereals, pulses, fruits and vegetables on dry-weight basis with 60-70 per cent digestibility. Mushrooms contain all essential amino acids and are good source of vitamin B, C, D and K and contain appreciable quantities of niacin, pantothenic acid, riboflavin, nicotinic acid and minerals like calcium, phosphorus and potash. It has been established that 100-200 g (dry weight) of mushrooms are sufficient to maintain nutritional balance in human being.

Three types of mushrooms are cultivated commercially in India and they are *Agaricus bisporus, Volvoriella spp.* and *Pleurotus spp.,* commonly known as European button, paddy straw and oyster mushroom. Mushroom cultivation requires high degree of humidity and therefore its cultivation in conjunction with aquaculture has tremendous scope. Method of cultivation involves use of dried paddy-straw chopped into 1.2 cm/bits, soaked in water overnight. Excess water is drained off. Horsegram powder (8 g/kg straw) and spawn (30 g/kg straw) is added and mixed with wet straw in alternating layers. Perforated polythene bags are filled with substrate and kept in room at 21°-35°C with required light and ventilation. The mycelial growth is seen in about 11-14 days which penetrates substrates in bags. Polythene bags are cut open at this stage without disturbing bed. Water is sprayed twice a day. In a few days mushroom crop becomes ready for harvest.

The paddy-straw after mushroom cultivation becomes rich in protein, organic nutrients and other matter. This supplemented nutritive value of used paddy-straw after harvest of the mushroom is utilized for cattle feeding. It has been recorded that such cattle feed enhances milk production. In turn, excreta of cattle is recycled in fish-ponds for enhancing pond productivity through detritus food-chain.

Seri–Fish System

Sericulture is an agro-industry and plays an eminent role in rural economy of India. Over 3 million people are employed in various fields of sericulture. The industry provides ample work for women-folk in rural area in rearing silkworm. The waste products from sericulture practices like silkworm pupae, faeces and wastewater from processing facilities could be used as nutrient input in fish-ponds. In this integration, mulberry is the producer; silkworm is the first consumer while fish is the secondary consumer, ingesting silkworm faeces directly. Inorganic nutrients in the silkworm faeces are utilized by phytoplankton, and heterotrophic bacteria are in turn consumed by filter-feeding fish, either directly or indirectly. Left-over feeds and fish faeces are decomposed by hydro-microbes releasing inorganic nutrients. The optimum range of temperature and humidity is 15-32°C and 50-90 per cent, for successful cultivation of silkworm.

In seri-fish system energy passes through complex food-web of the dyke-pond system and undergoes a series of exchange as it flows among sub-systems. It forms a complex food which is extended in various forms and via various pathways of the system. Some energy escapes, such as those stored in silkworm cocoons or in the fish, are of economic value. Others, like losses in the form of radiation is the energy source that drives dyke-pond system. This energy enters system via 3 pathways. Absorption by dyke-crops converts energy into chemical energy during photosynthesis. Absorption by phytoplanktons in the pond converts chemical energy via photosynthesis and is direct input to pond. Chemical energy stored in plant material and waste production is used as fish feed and pond fertilizer. The silkworm sub-system provides energy linkage between mulberry and pond sub-system. It absorbs stored energy in harvested mulberry leaves and waste material obtained in silkworm rearing enters fish-pond as a mixture of mulberry leaves and silkworm excrement. In other words, 75 per cent of the mulberry-leaves, supplied is consumed by silkworm. Together with silkworm excrement, the remaining 25 per cent of unconsumed leaf debris is dumped into pond. Mulberry dykes yield leaves at 30 tonnes/ha/year, when fed to silkworm 16-20 tonnes of waste is produced, in which stored energy is 66 per cent of that supplied to silkworm. The energy intake by fish accounts for only 32 per cent of the total input. In 1 ha mulberry-pond system, 50 per cent of area is kept for dyke and 50 per cent is water area. Of the former, 0.45 ha is planted with mulberry and remaining 0.05 ha is used for crop production. During winter, vegetables are inter-planted with mulberry. In this method, 30 tonnes of mulberry-leaves/ha, 3.75 tonnes of vegetables/ha, and several tonnes of crop can be produced. Waste of vegetables which account 50 per cent is fed to fish, while remaining 50 per cent is consumed by human-beings.

Chapter 12

Nutrient Dynamics in Integrated Fish Farming Systems

Nutrient dynamics in any aquatic ecosystem deals with the study of major plant nutrients as they move from one compartment to another compartment and from one phase to another. The source of available nutrients in any aquatic system is of two types (*i*) internal-autochthonous (*ii*) external – autochthonous. Internal sources are derived from either sedimentary phase or by natural fixation processes (this applies only to the major plant nutrient nitrogen as it is generally derived from nitrogen fixers utilizing the atmospheric nitrogen and which is very common in freshwater ecosystem). The major plant nutrients that stimulate phytoplankton growth in any aquatic ecosystem include carbon, nitrogen, potassium and phosphorus. In this, potassium is available in high quantities in many aquatic environments and therefore, dynamics of potassium does not warrant any great care as it is in terrestrial environments. To maintain adequate phytoplankton as natural food to support desirable fish yield, fertilization by adding nutrients either organic or inorganic forms is imperative. Theoretically pond fertilization should be based on Liebig's law of minimum, which states that biological production is limited by the nutrients in least supply. Under limited nutrients supply, the rate of nutrients uptake by phytoplankton is concentration dependent and the total phytoplankton production is directly proportional to the initial concentration of limiting nutrients. However, in practice, optimal pond fertilization is an extremely complex matter due to the dynamics of intrinsic and extrinsic factors in pond ecosystems. For example, the difference in nutrient requirements of individual algal species leads to a need to provide a balanced supply of major nutrients.

The Stoichiometric composition of freshwater phytoplankton can be given as $C_{106'}$ $H_{180'}$ $O_{45'}$ N_{16} P_{1}. Nitrogen, phosphorous and occasionally carbon are the most common limiting nutrients to phytoplankton in fishponds. As per the above composition C: N: P=106:16:1. Under optimal growth conditions, the average nutrient composition of phytoplankton biomass are approximately 45 to 50 per cent carbon, 8 to 10 per cent nitrogen and 1 per cent phosphorus, giving a typical C: N: P ratio of about 50:10:1. The minimal concentration of a given nutrient required to meet the optimal growth is referred as "critical concentration". Algae are capable of continuously utilizing the nutrients above this concentration and deposit the surplus nutrients in their cells without concomitant growth, which is termed "luxury consumption". This phenomenon is particularly well known for phosphorus uptake. The standing algal crop in fishponds is continuously and partially consumed by filter feeding organisms and fish like herbivores in the pond ecosystem. Thus the amount of nutrients required is only to compensate for algal growth and the losses through grazing and sinking. Furthermore, only a fraction of the nutrients is ingested/assimilated by fish and other organisms and a large portion is recycled back to ponds as wastes.

The nitrogen, phosphorus ratio is an important consideration in pond fertilization,. Although the typical N: P ratio in algal biomass is roughly 10:1, the fertilizer composition for pond input deviates considerably from this ratio because of the markedly different natural dynamics of nitrogen and phosphorus in ponds. Nitrogen is considered to be less limited in our aquatic environment because of the input from natural nitrogen fixation through biological and atmospheric processes. Earlier nitrogen inputs were thought to be sufficient for low levels of pond production and the nitrogen fertilization was considered as unnecessary. In contrast, phosphorus is limited in most aquatic ecosystems, as its source depends primarily on weathering of phosphorus rocks, which are not ubiquitous in the lithosphere. Furthermore, the availability of dissolved phosphorus in aquatic ecosystem is curtailed by its rapid reaction with cations like iron, calcium, magnesium and aluminium forming relatively insoluble precipitates, which are temporarily or permanently unavailable to biota. Phosphorus present in the water column is also removed by pond mud's, which strongly adsorb phosphorus. Mud adsorption capacity of phosphorus is linearly correlated to clay content of the sediments. Thus the amount of phosphorous required for pond fertilization is influenced by the type of bottom soils and their percentage of phosphorus saturation factors, which are often related to the fertilization history of the ponds. Surplus inputs of phosphorus fertilizer in ponds often result in phytoplankton "luxury consumption" forming intracellular condensed inorganic polyphosphates. In case of deficient nutrient supply from ambient water, those reserves can serve as a potential source of nutrient for further growth through phosphate hydrolysis. In contrast, excess nitrogen inputs may lead to ammonia toxicity to fish. When the appropriate N:P ratio for a fertilization regime in a pond is not known, a surplus of phosphorus is safer than a surplus of nitrogen. The ammonia concentration in pond water should not exceed 0.5 mg/litre.

The nutrients contents of animal manure may vary overtime and the nutrient availability to phytoplankton growth remains unclear. The rate of nutrient released from animal manure over time is a key factor to decide the frequency and the amount of manure required to fertilize the ponds. The N and P ratio in faecal wastes of most animals remains consistently around 2 but N in urine is markedly higher than that in faeces. Chicken faeces contain relatively higher N and P than those from large ruminants eg: Buffaloes and white cattle. In general, the moisture and nutrient contents of manure may vary considerably, depending on factors such as the diet purity and treatment of manure and duration and conditions of storage. Fresh poultry droppings contain twice as much nitrogen as farmyard manure and are much richer in P. The uric acid in fresh droppings is decomposed quickly by microorganisms, releasing ammonia, which is easily lost upon exposure of manure to air. The quality of manures depend on their composition as they are often mixed with decomposed animal manure, plant residues, sawdust, and lime. In many instances animal manures may contain a significant amount of spilled animal feed, which contributes directly as fish diet when applied to ponds. This is particularly obvious in a poultry fish integrated system. Around 51 to 57 per cent of the total nitrogen in dry chicken manure was released as dissolved inorganic nitrogen to water during the first three days, while it takes twenty days to release 68 to 73 per cent of total phosphorus as dissolved inorganic phosphorus. In the case of duck manure, almost all soluble nitrogen and phosphorus are released into the pond water within four days of manure application. Around 90 per cent of the total nitrogen is released as dissolved inorganic nitrogen in case of buffalo manure within three days and whereas the dissolved inorganic phosphorus reached a maximum concentration after four days, which is equivalent to 35 per cent of the initial total phosphorus content of the manure. Decomposition of buffalo manure results in dark brown water colour and high concentration of suspended solids, which reduces Secchi disc transparency.

Animal manures are generally low in N:P ratio (<3). Such a low N:P ratio is not optimal for phytoplankton growth. Thus fish pond can be fertilized at a reduced manure rate and supplemented with inorganic nitrogen fertilizer to provide adequate nitrogen and avoid oxygen depletion.

Pond sediment receives a great deal of organic matter originating from the water column. Sedimentation of particulate organic matter and detritus formed in the water column are the major sources of organic matter in fish pond sediment. Organic matter arriving at pond sediment will exit from the nutrients dynamics in the water column, while precipitating in the ecological processes in sediment, which is considered to be very slow. Thus, the sediment or mud in fish pond is regard to be an "energy sink" in the fish pond ecosystem. Bioturbation by benthivorous and detritivorous fishes facilitates the penetration of dissolved oxygen into the sediment pore water and enhances the mineralization of organic phosphorus in the sediment. Further phosphate ions associated with sediment particles is liberated into the water column when sediment particles are resuspended by the activities of benthivorous and detritivorus fishes. The sediments of any aquatic ecosystem are considered to be phosphorus sink. However, in a fish pond stocked with benthivorous fishes, sedimentary phosphorus dynamics is greatly enhanced. In fish ponds, there exists several coupled, positive feed back processes which drive the liberation of phosphorus from the sediment to the water column and reduce the phosphorus content of the sediment compartment to a lower level than that observed in some eutrophic lakes.

The nitrogen in the pond ecosystem occurs as different chemical species in water and sediment compartments. Ammonia, nitrite and nitrates are the different fractions of soluble inorganic nitrogen. The soluble organic nitrogen consists of amino acids, enzymes, peptide, uric acid, urea etc., derived from the wastes of living organisms and those derived from the decaying and decomposed organic matter. The particulate fractions of nitrogen in water are of two types. (i) Particulate inorganic nitrogen, (ii) Particulate organic nitrogen. The particulate inorganic nitrogen occurs as nutrient adsorped mineral or clay particles. The living organisms of different sizes such as bacteria to nekton community and the particulate detrital matter forms the particulate organic nitrogen fraction. The sediment compartment in the pond ecosystem possesses four different types of chemical species of nitrogen such as (*i*) Exchangeable ammonia-nitrogen (*ii*) Exchangeable nitrite –nitrogen (*iii*) Exchangeable nitrate- nitrogen and (*iv*) Sedimentary organic nitrogen.

Biosynthesis of protein in any fish culture system depends entirely upon the autochthonous and allochthonous nitrogen availability in the system. When organic nitrogen compounds are hydrolysed and catabolised, nitrogen is liberated as ammonia. Ammonia forms the most preferential form of nitrogenous nutrient in any aquatic ecosystem and urea ranks next to it followed by nitrate. If the ammonia-nitrogen concentration in the water column is above 1-2µg at-N/1 and that would highly inhibit nitrate uptake by phytoplankton. Under normal conditions, in freshwater fish culture ponds even 300 per cent reduction of ammonia concentration is possible due to peak photosynthetic activity. Accordingly, in noon hours with very high rate of primary production, nitrate assimilation by photoplankton is more likely to happen as the ammonia level may fall below 2 µg at-N/1. Sediments play a vital role in the quantitative transformation of ammonia to nitrate. However, the very long duration of time taken for the oxidation of ammonia to nitrite could not be explained in the light of available literature and it was more striking to observe that in the field conditions nitrite always occurred in trace level. In sediments, the dominance of nitrate does not pose any problem either to the system or to the fishes under culture. But it will sharply counteract hydrogen sulphide production by triggering denitrification. Thus, higher availability of nitrate in sediments would safeguard the health and ecological balance of these ponds from any occasional onset of anoxic condition.

The different chemical forms of phosphorus available in the water medium is as follows:

1. Soluble inorganic phosphorus (SIP)
2. Soluble organic phosphorus (SOP)
3. Polyphosphate (PyP)
4. Particulate phosphorus (PP)

Soluble inorganic phosphorus is conventionally called, as "Soluble orthophosphate" and is known to be directly utilized by phytoplankton. The colloidal substance with a molecular weight less than 10 and commonly occurring in fish culture ponds play a vital role in the dynamics of phosphorus. The phosphorus present in the water column is not biologically assimilated but in turn precipitated as apatite contributing to more than 80 per cent of the total sedimentary phosphorus, which cannot be recycled in the system. This could be averted by the use of organic manure, which would check the possibility of apatite production to some extent by solubilizing the biogenic lime produced during active photosynthesis. In any fish culture receiving more load of organic manuring, the soluble organic phosphorus could dominate over other soluble chemical species of phosphorus in the water medium. Intra and extra cellular phosphatase enzymes of algae and bacteria are capable of producing SIP by hydrolyzing soluble organic phosphorus. Under phosphorus limiting condition, the specific activity of these enzymes has been reported to increase tremendously. Polyphosphates occur in natural waters and are primarily due to domestic pollution. High and low molecular weight polyphosphates are synthesized within the body of many algal species. In any fish culture ponds, freshly formed apatite and phosphate chemo-adsorbed on biogenic lime contribute to significant percentages of particulate phosphorus. In addition to this, different chemical (due to the agitation caused by winds and bioturbation) also contribute to significant percentages of particulate phosphorus present in the water column.

The sedimentary phosphorus comprises of the following different chemical species as its components.

1. Neutral hydrolysable phosphorus (NH-P)
2. Phosphorus bound to sedimentary carbonates (CO_3-P)
3. Phosphorus bound to sedimentary oxides of iron and manganese (O-Fe & Mn-P)
4. Phosphorus bound to sedimentary iron and aluminium (Fe & Al-P)
5. Apatite phosphorus (A-P)
6. Phosphorus bound to sedimentary organic matter (Po)
7. Residual phosphorus (R-P)

Phosphorus content of water in aquatic ecosystems is highly influenced by the bottom sediments. Sediments act as a net sink for phosphorus due to sedimentation of suspended matter. Sediments act as a transient source of phosphate and in turn decide the maintenance of water quality in aquatic ecosystems. The different chemical species of sedimentary phosphorus regulate the phosphorus content of the water through many physico-chemical and biological factors.

NH-P has been reported to give a measure of the "Surplus phosphorus" of the phytoplankton that settle over the bottom sediments. This fraction of sedimentary phosphorus also represents easily, hydrolysable organic phosphorus and adsorbed inorganic and polyphosphates of the sediments. Significant quantity of nutrient element should be associated with sedimentary carbonates and such element would be more susceptible to marked changes in pH. Adsorption of phosphate on biogenic lime and decomposition of organic matter in a freshwater fish pond would highly decide the availability of this fraction of phosphorus in sediments.

O-Fe & Mn-P representing phosphorus scavenged by iron and manganese oxides and is more affected by anoxic condition. Thus, the availability of this fraction indirectly explains the status of health of the aquatic ecosystem. Po has negative correlation with non-residual sedimentary phosphorus and functioned negatively with respect to the accumulation of other labile fraction of phosphorus in the bottom sediment. A-P accumulates in the bottom sediment due to the co-precipitation of phosphate with calcium carbonate into crystalline apatite and which cannot be biologically utilized and easily regenerated. Fe & Al-P is an easily exchangeable fraction in freshwater system. The release of phosphorus bound to iron and aluminium is governed by factors like reducing conditions and heavy biological uptake. This exchangeable fraction reflects the extent of adsorption effected between the sediment and water by redox potential. R-P did not exhibit any marked time bound variation in the distribution and this fraction chiefly reflects the quantity of phosphorus bound in the crystal structure of primary and secondary minerals and does not undergo any change due to the physico-chemical and biological factors normally operative in any aquatic ecosystem. Thus, the applied (allochthonously) fertilizer and manures along with the derived (autochthonously) nutrients in any aquatic ecosystem are undergoing complex dynamic processes.

Chapter 13
Chemical Composition of Animal Waste

The nutrients in animal wastes are considered as valuable resources. The proportion of input feed nutrients remaining in the wastes accounts to 72-79 per cent of the nitrogen, 61-87 per cent of the phosphorus, and 82-82 per cent of the potassium and this potential value for fish culture is being well appreciated (Edwards, 1980). Waste output in the form of urine and faeces varies considerably in quantity and quality. Feed (amount and composition) and water intake, climate and management methods will also affect the value of waste to fish culture. In general it would be expected that animals fed with high quality feed would have more nutrients in the wastes. Similarly, livestock fed with large quantities of feed would also have more nutrients since the faster passage through the gut would prevent efficient digestion and absorption (Edwards, 1982). The quality and quantity of wastes from fattening units is also different from those from breeding animals.

The relative distribution of nutrients in solid and liquid wastes can also vary. Thus, higher levels of nitrogen (N) and potassium (K) have been found in urine (comprising 40 per cent by weight of total waste excretion) than in faeces. A high phosphorus (P) content can be found in the faeces of animals, except pigs, which have considerable phosphorus in their urine (Delmendo, 1980).

Table 12: Farm Animal Waste Output

	Pigs	*Hens*	*Ducks*	*Cattle*	*Horses*	*Sheep*
Animal weight (kg)	55	2	3	500	380	30
kg wet waste/ animal/day	8	0.7	1.0	30	24	2.1
Per cent faeces	45	–	–	70	70	66
Per cent urine	53	–	–	30	30	34

Source: Meynell, 1982.

The livestock waste includes dung, urine, feed refuse, litter/bedding and washed. These wastes are not utilized properly and hence more attention should be given for getting maximum economic benefits. Among the livestock waste dung is the widely used materials for manure, fuel and as a feed for fishes. According to the survey conducted by the National Council of Economic Research, the annual wet dung production in India is estimated at 1335 million tonnes. Nearly 60-70 per cent of this is used in the form of 'Pathis' (dung cakes) as fuel, while the entire urine goes waste. Ranjan and Pathak (1979) reported that 92 per cent of these waste is contributed by the dung of cattle and buffaloes as shown below:

Table 13: Production of Livestock Waste per annum in India (Million tons on dry matter)

Livestock waste	*Quantity*	*Relative per cent*
Cattle and buffalo dung	267.0	92.6
Sheep and goat faeces	15.5	5.4
Pig	1.9	0.7
Poultry droppings	1.3	0.4
Other waste	2.5	0.9
Total		288.2

The average production of wastes from different categories of animals on fresh matter basis is furnished in Table 14.

Cattle contributes to the bulk of the farmyard manure. Besides cattle, sheep, goat and swine also contribute to some extent. In the last few decades the poultry industry has grown tremendously and its contribution is also substantial. An adult pullet consumes 45kg of balanced feed containing all essential ingredients to maintain life and also maintain production. The diet fed to birds is digested to the extent of 70-80 per cent and the rest is excreted in faeces and hence it contains 20-30 per cent indigested food and may also contains micronutrient added to the faeces due to metabolic origin and issue waste.

Generally, 100 laying hens will excrete 10 – 12 kg of wet droppings per day having 70 – 75 per cent moisture depending on season. On dry matter basis 2.5-3.0 kg manure is made available per day/100 birds. The annual available manure per bird is about 10 kg poultry manure on dry matter basis.

Table 14: Average Production of Waste per day on Fresh Matter Basis

Waste Property	*Symbol*	*Unit*	*Pigs*	*Layer hens*	*Sheep*	*Dairy Cattle*
Excreted Wet Waste	TWW	Kg/d of APU*	5.1	6.6	3.6	9.4
Urine	U	% of TWW	45.0	0	50.0	31.0
Total Solids	TTS	% of TWW	13.5	25.3	29.7	9.3
		Kg/d of APU	0.69	1.67	1.07	0.89
Moisture Content	TMC	Per cent of TWW	86.5	74.7	70.3	90.7
Total Volatile	TVS	Per cent of TTS	82.4	72.8	84.7	80.3
Solids		Kg/d of APU	0.57	1.22	0.91	0.72

*: APU-Animal Production unit is equal to 100kg of total live weight of the animal.

The comparative chemical composition of excreta of different animals are presented in the Table 15.

Table 15: Chemical Composition of Excreta from Different Kind of Livestock (in percentage)

Material	*DM*	*Volatile Organic Matter*	*Fat*	*Crude Protein*	*Crude Fibre*	*Ash*
Cow dung	18	84	2.4	9.2	21	16
Buffalo dung	19	83	3.0	9.9	19	17
Sheep droppings	32	81	3.8	17.1	13	19
Goat droppings	32	78	3.2	12.5	13	22
Piggery waste	45	72	4.5	19.4	26	28
Poultry droppings	47	65	1.5	26.9	19	34

In a study by Saikia *et al.* (1988), the poultry manure collected from a layer house and cow dung collected from milking yard of Jersey cows were dried in sunshine and analysed. The chemical composition of cage poultry manure and cow dung manure on dry matter basis is presented in Table 16:

Table 16: Chemical Composition of Cage Poultry Manure and Cow Dung Manure (Per cent on dry matter basis)

Nutrients	*Dried Poultry Manure*	*Dried Cow Dung Manure*
Crude Protein	22.50	14.05
Ether Extract	1.68	2.40
Crude Fibre	9.56	22.02
Nitrogen Free Extract	46.04	38.28
Total Ash	20.22	23.25
Insoluble ash	9.50	12.23
Calcium	3.43	0.89
Phosphorus	1.58	1.89

The livestock waste are measured in terms of its Nitrogen (N), Phosphorus (P) and Potassium (K) content of farm yard manure and broiler raised under deep litter system were furnished in Table 17:

Table 17: NPK Content of Farm Yard Manure and Broiler Deep Litter

Class of Livestock	*Nitrogen in Terms of Urea%*	*Phosphorus in Terms of Super-phosphate%*	*Potash in terms of Muriate of Potash%*
Farm Yard Manure	0.90	0.60	1.10
Broiler I st week	1.52	1.03	1.06
Broiler 4th week	3.49	2.21	1.76
Broiler 8th week	3.65	1.83	1.66

In another study, Mallikeswaran *et al.* (1984) reported the composition of various animal wastes and fertilizers which is presented in Table 18.:

Table 18: Composition of Various Animal Wastes and Fertilizers

Sl.No.	*Animal Waste*	*Nitrogen (%)*	*Phos-phorus (%)*	*Potash (%)*
1.	Cage layer manure wet 63% moisture	1.3	0.8	0.7
2.	Cage layer manure Dry 18% moisture	3.2	1.7	1.2
3.	One year old deep litter manure	3.0	2.0	2.0
4.	Farm Yard manure wet	0.5	0.2	0.5
5.	Ammonium Sulphate	20.0	–	–
6.	Urea	45.0	–	–
7.	Super Phosphate	–	22.9	–
8.	Sulphate of Potash	–	–	48.0

Table 19: Nutritive values of different animal excreta

Animal	*Excreta*	*Moisture (%)*	*Organic Matter (%)*	*Nitrogen (%)*	*Phosphorus (P_2O_5) (%)*	*Potash (K_2O) (%)*
Cattle	Faeces	80-85	14.0	0.3	0.2	0.1
	Urine	92-95	2.3	1.0	0.1	1.4
Pig	Faeces	85	15	0.6	0.5	0.4
	Urine	97	2.5	0.4	0.1	0.7
Chicken	Faeces	78	25.5	1.4	0.8	0.6
Duck	Faeces	81	26.2	0.9	0.4	0.6
Rabbit	Faeces	10	37	2.0	1.3	1.2
Goat	Faeces	10	–	2.7	1.7	2.9

Source: Integrated fish farming, NACA Technical Manual 7.

The animal waste can yield still higher economics if a part of it can be used for feeding back to the animals. But before the waste is used as a feed supplement for the animals one ought to know

1. The nutritive value,
2. Its safety on animals and
3. The possible residual effects in the animal products to be consumed.

These aspects are important because the utilization of waste from large animal feed lots may create various health problems for both animals, as well as human beings and may also cause environmental pollution.

Chapter 14

Economics of Various Integrated Fish Farming Practices

A) Economics of Fish-Cattle Integrated Farming

(i)	**Capital investment**		
1.	Cost of cross-bred cows 5 No. @ Rs.30,000/cow	:	Rs.1,50,000
2.	Cost of construction of cow-shed (5x50 sq.ft/cow x Rs.300/sq.ft)	:	Rs. 75,000
3.	Cost of equipments Rs.600/cow	:	Rs. 3,000
4.	Cost of pump set	:	Rs. 20,000
5.	Cost of inlet-outlet structures	:	Rs. 10,000
	Total capital investment	:	Rs.2,58,000

(ii) Costs

(a) Fixed costs

1.	Depreciation on capital items @ 10% (except for cows)	:	Rs. 10,800
2.	Interest on capital @ 15%	:	Rs. 38,700
3.	Repairs and upkeep	:	Rs. 5,000
4.	Insurance premium @ 6% on the value of cows	:	Rs. 9,000
5.	Rent or rental value of land per hectare	:	Rs. 10,000
6.	Wages to permanent labour 12 months (at Rs.1500/labour/month)	:	Rs. 18,000
	Total fixed costs	:	Rs. 91,500

(b) Variable costs

1.	Pond preparation cost (Size – 1 ha)	:	Rs. 5,000
2.	Cost of fish seeds 10,000 nos @ Rs.500/1000	:	Rs. 5,000
3.	Cost of cattle feed 4 kg/cow/day for 300 days and 2kg/cow/day for 65 days 5x1330 kg @ Rs.12/kg	:	Rs. 79,800
4.	Cost of fodder grass 15 kg/cow/day 5x3475 kg @ Rs.1/kg	:	Rs. 27,375
5.	Veterinary expenses @ Rs.300/cow/yr x 5	:	Rs. 1,500
6.	Electricity expenses @ Rs.1 x 12	:	Rs. 1,200
7.	Wages for fish harvesting for one hectare	:	Rs. 5,000
	Total variable cost	:	Rs.1,24,875
	Total costs (a+b)	**:**	**Rs.2,16,375**

(c) Net returns

1.	By sale of fish (4000 kg @ Rs.50/kg)	:	Rs.2,00,000
2.	By sale of milk (10 litres/cow/day x 300 days x 5 x Rs.15/litre)	:	Rs.2,25,000
	Total returns	:	Rs.4,25,000
	Net returns	**:**	**Rs.2,08,625**

(B) Economics of Fish-Pig Integrated Farming

(i)	**Capital investment**		
1.	Cost of construction of pigsty 35x16 sq.ft/pig @ Rs.300/sq.ft	:	Rs.1,68,000
2.	Cost of equipments @ Rs.150/pig	:	Rs. 5,250
3.	Cost of pump set	:	Rs. 20,000
4.	Cost of inlet-outlet structures	:	Rs. 10,000
	Total capital investment	:	Rs.2,03,250
(ii)	**Costs**		
(a)	*Fixed costs*		
1.	Depreciation on capital items @ 10%	:	Rs. 20,325
2.	Interest on capital @ 15%	:	Rs. 30,488
3.	Repairs and upkeep	:	Rs. 5,000
4.	Insurance premium @ 6% on the value of pigs	:	Rs. 378
5.	Rent or rental value of land per ha	:	Rs. 10,000
6.	Wages to permanent labour-at Rs.1500/month x 12	:	Rs. 18,000
	Total fixed costs	:	Rs. 84,191
(b)	*Variable costs*		
1.	Pond preparation cost	:	Rs. 5,000
2.	Cost of fish seeds 10,000 nos @ Rs.500/1000	:	Rs. 5,000
3.	Cost of piglets (8 kg size 35 nos./cycle x 2 cycles (@ Rs.250/piglet)	:	Rs. 17,500
4.	Cost of pig mash @ 400 kg/animal/35 animals for 2 cycles @ Rs.1/kg	:	Rs. 28,000
5.	Veterinary expenses @ Rs.50/pig/x 70	:	Rs. 3,500
6.	Electricity expenses @ Rs.200/month x 12	:	Rs. 2,400
7.	Wages for fish harvesting for one hectare	:	Rs. 5,000
	Total variable cost	:	Rs. 66,400
	Total costs (a+b)	**:**	**Rs.1,50,591**

(c) Net returns

1.	By sale of fish (4000 kg @ Rs.50/kg)	:	Rs.2,00,000
2.	By sale of pig (4200 kg at Rs.50/kg)	:	Rs.2,10,000
	Total returns	:	Rs.4,10,000
	Net returns	:	**Rs.2,59,409**

(C) Economics of Fish-Poultry Integrated Farming

(i) Capital investment

1.	Cost of construction of poultry shed 500x1 sq.ft @ Rs.200/sq.ft	:	Rs.1,00,000
2.	Electrical installations and overhead water provision 500xRs.20/bird	:	Rs. 10,000
3.	Cost of pump set	:	Rs. 20,000
4.	Cost of inlet-outlet structures	:	Rs. 10,000
	Total capital investment	:	Rs.1,40,000

(ii) Costs

(a) Fixed costs

1.	Depreciation on capital items @ 10%	:	Rs. 14,000
2.	Interest on capital @ 15%	:	Rs. 21,000
3.	Repairs and upkeep	:	Rs. 5,000
4.	Insurance premium 500 x Rs.2.00/bird	:	Rs. 1,050
5.	Rent or rental value of land per hectare	:	Rs. 10,000
6.	Wages to permanent labour-at Rs.1500/month x 12	:	Rs. 18,000
	Total fixed costs	:	Rs. 69,050

(b) Variable costs

1.	Pond preparation cost	:	Rs. 5,000
2.	Cost of fish seeds 10,000 nos. @ Rs.500/1000	:	Rs. 5,000
3.	Cost of layer birds 500xRs.90/bird	:	Rs. 45,000
4.	Cost of poultry feed 500 x 40 kg/bird @ Rs.12/kg	:	Rs.2,40,000
5.	Veterinary and electricity expenses 500 x Rs.5/bird	:	Rs. 2,500
6.	Wages for fish harvesting for one hectare	:	Rs. 5,000
	Total variable cost	:	Rs.3,02,500
	Total costs (a+b)	**:**	**Rs.3,71,550**

(c) Net returns

1.	By sale of fish (4000 kg @ Rs.50/kg)	:	Rs.2,00,000
2.	By sale of egg (450x250 eggs/bird at Re.2/egg)	:	Rs.2,25,000
3.	By sale of culled birds (450 x 2 kg/bird @ Rs.80/kg)	:	Rs. 72,000
	Total returns	:	Rs.4,97,000
	Net returns	**:**	**Rs.1,25,450**

(D) Economics of Fish-Duck Integrated Farming

(i) Capital investment

1.	Cost of construction of duck pen 200x1 sq.ft/duck @ Rs.100/sq.ft	:	Rs. 20,000
2.	Cost of pump set	:	Rs. 20,000
3.	Cost of inlet-outlet structures	:	Rs. 10,000
	Total capital investment	:	Rs. 50,000

(ii) Costs

(a) Fixed costs

1.	Depreciation on capital items @ 10%	:	Rs. 5,000
2.	Interest on capital @ 15%	:	Rs. 7,500

3.	Repairs and upkeep	:	Rs. 5,000
4.	Insurance premium @ Rs.1.50/bird	:	Rs. 300
5.	Rent or rental value of land per ha	:	Rs. 10,000
6.	Wages to permanent labour-at Rs.1500/month x 12	:	Rs. 18,000
	Total fixed costs	:	Rs. 45,800
(b)	*Variable costs*		
1.	Pond preparation cost	:	Rs. 5,000
2.	Cost of fish seeds 10,000 nos @ Rs.500/1000	:	Rs. 5,000
3.	Cost of ducks 200 nos @ Rs.100/duck	:	Rs. 20,000
4.	Cost of duck feed 200 x 30 kg/duck @ Rs.10/kg	:	Rs. 60,000
5.	Veterinary and electricity expenses 200 x Rs.5/duck	:	Rs. 1,000
6.	Wages for fish harvesting for one hectare	:	Rs. 5,000
	Total variable cost	:	Rs. 96,000
	Total costs (a+b)	**:**	**Rs.1,41,800**
(c)	*Net returns*		
1.	By sale of fish (4000 kg @ Rs.50/kg)	:	Rs.2,00,000
2.	By sale of duck egg (180x120 eggs/bird at Rs.3/egg)	:	Rs. 64,800
3.	By sale of culled ducks (100 x 1.5 kg/bird @ Rs.50/kg)	:	Rs. 13,500
	Total returns	:	Rs.2,78,300
	Net returns	**:**	**Rs.1,36,500**

Chapter 15
Constraints in Integrated Fish Farming

In India most of the integrated aquafarms are operated in a traditional way without proper planning, application of available technology and management techniques. The farmers mostly use their personal experience for farming. Another constraint is marketing of the farm produce except in places where there are established markets. Due to lack of technical knowledge and inadequate experience farmers incur heavy losses due to disease outbreaks. The input dealers supplying animal feeds and veterinary medicines generally offer extension services to farmers, but due to lack of proper roads and remote, distance, this practice is also being limited. Farmers usually sell their produce to middleman at low price because of lack of working capital.

A sustainable technology is the need of the hour for higher production from existing agricultural land and water source. In this regard, integrated farming offers a possible solution and holds a great promise and potential for augmenting production, betterment of rural economy and employment generation, thus finally enhancing the socio-economic status of weaker rural community.

The advantages of effective wastes collection from intensive animal production are imperative. Factors acting against such intensive production may pose constraints for any development of integrated animal/fish systems. The absence of good stock, concentrate feeds and management skills may also reduce the potential benefits of such systems.

Cattle and buffalo manure is widely used as a fuel in India and chicken manure commands a high price in many places because of its value as a fertilizer. In such case, competing economic benefits must be considered carefully, and the costs of reduced availability for other purposes must be included in proper assessment of integrating with aquaculture.

A reluctance by farmers to handle animal wastes and then to consume the cultured fish, by utilising the wastes can be a severe constraint on the successful promotion of integrated aquaculture. Direct addition of wastes by housing the stock over or near the pond may reduce the necessity to handle fresh wastes. The latter problem may be more difficult to resolve however; some racial and ethnic groups are more averse to the practice of eating waste-fed fish than others. This problem can some how solved if the produce are sold far away from the farm area.

Land, water or labour availabilities may be more limiting than food or fish. As a result intensification of fish production may increase to the point where the physical integration with livestock becomes uneconomic. Thus the characteristics and limitations of a livestock waste-fed pond may not give a predictable and marketable enough fish harvest.

The dangers of fish acting as vectors for human pathogens are not yet defined clearly, although there are some studies which indicate the risks. Velasquez (1980) has reviewed the diseases potentially communicable to man via fish and the water of animal/fish systems in Philippines. Helminth infections are identified as being particularly dangerous, especially since the encysted form within the fish tissue can survive customary methods of preparation and preservation. In addition to a possible role in the spread of *Salmonella*, integrated fish systems may be implicated as the passive vectors of other bacterial diseases. Hubbert (1983) reported increased appearance of *Edwardsiella* spp. in the guts of carp fed digested cow manure, and noted that this type of bacteria has been associated with gastric infections in man, especially in the tropics.

Certain types of wastes are more likely to adversely overload or underload the pond system, causing a reduction in fish yield. Confinement methods for livestock often involve the use of bedding materials that alter considerably the available wastes for aquaculture. Nutrient concentrations, and the proportion of fibrous materials would typically change. Some materials, such as rice husk, which is often used as a bedding litter material for broiler chickens, are known for their slow decomposition. Prolonged storage of such materials is thus required before their utilisation as pond inputs, which will also tend to reduce the waste's value.

In other circumstances, the mixing of easily biodegradable litter and its rapid aerobic composting, may make greater quantities of quality input materials available. The higher nutrients quality of some wastes, may result in easy overloading of a pond system, with deleterious consequences for fish.

While land preparation for paddy poses at times difficulty for fish culture, fields without proper water structures either suffer badly with flooding and consequently loss of fish or with water shortage restricting the period of fish culture. High temperature and turbidity of the paddy fields adversely affect fish life.

Infestation through the water supply of undesirable species of fish becomes competitors with commercially important fishes for oxygen, space and food organism. There are numerous kinds of insects such as Corixa, Dysticus, Belostoma, Hydrometra, Gerris etc. in the paddy-field which predate on fish larvae and fry whereas leech, frog, birds, snakes and otters prey upon fish in paddy-fields.

Fish cultivation along with rice is affected by the extensive usage of pesticide. While the International Rice Research Institute (IRRI), Philippines is evolving certain strains of rice which are highly disease-resistant, biological control of pest and predators of paddy are also being vigorously investigated in many places. However, presently, the best way to apply the chemicals is either to drive the fish before the application into the trenches or to increase the water level of the field and try to spray the chemicals on the leaves so that little amount of such chemicals drop into the water and their concentration gets diluted.

While increased use of organic manures and crop residues is in practice in puts such as fish seed, feed and breed–the three major input components for successful crop-fish-livestock production systems are by and large beyond the reach of poor farmers. This needs to be seriously viewed by all concerned.

Low investment capacity of the farmer, credit constraints, untimely supply of inputs, non-availability of proper agricultural implements; leasing policy, multiple ownership, absentee landlordism and marketing facilities are the common socio-economic constraints responsible for slow development of such integrated systems.

References

Agarwal, S.C., 1994. *A Handbook of Fish Farming*. Narendra Publication House, Delhi, 117 pp.

Ahilan, B. *et al.*, 2009. *Goldfish*. Daya Publishing House, New Delhi, 90 pp.

Ahilan, B. *et al.*, 2008. *Textbook of Aquariculture*. Daya Publishing House, New Delhi, 157 pp.

Boyd, C.E., 1995. *Bottom Soils, Sediment and Pond Aquaculture*. Chapman & Hall, New York, 348 pp.

David, C. and James, M., 1987. *A Guide to Integrated Warm Water Aquaculture*. Institute of Aquaculture Publications, University of Stirling, Scotland, 238 pp.

FAO, 2005. *Integrated Agriculture Aquaculture: A Primer*. Daya Publishing House, New Delhi, 149 pp.

Gopalakrishnan, C.A. and Mohanlal, G.M., 1993. *Livestock and Poultry Enterprises for Rural Development*. Vikas Publishing House, Pvt. Ltd., Delhi, 1096 pp.

Jagtap, H.S. *et al.*, 2009. *Practical Manual of Pisciculture and Aquarium Keeping*. Daya Publishing House, New Delhi, 201 pp.

Jagtap, H.S. *et al.*, 2009. *A Textbook of Pisciculture and Aquarium Keeping*. Daya Publishing House, New Delhi, 263 pp.

Jhingran, V.G., 1991. *Fish and Fisheries of India*. Hindustan Publishing Corporation, Delhi, 622 pp.

Mathias, J.A., Charles, A.T. and Baotong, H., 1994. *Integrated Fish Farming*. CRC Press, USA, 420 pp.

Overcash, M.R., Humemik, F.J. and Miner, J.R., 1983. *Livestock Waste Management,* Vols. I & II, CRC Press Inc., USA, 244 pp.

Pillay, T.V.R., 1993. *Aquaculture: Principles and Practices.* Fishing News Books, Farnham, Surrey, England, 515 pp.

Pillay, T.V.R. and Dill, Wm.A., 1976. *Advances on Aquaculture*. Fishing News Books, Farnham, Surrey, England, 653 pp.

Rath, R.K., 1993. *Freshwater Aquaculture*. Scientific Publishers, Jodhpur, 193 pp.

Santhanam, R., 2008. *Fisheries Science*. Daya Publishing House, New Delhi, 174 pp.

Santhanam, R., Sukumaran, N. and Natarajan, P., 1990. *A Manual of Freshwater Aquaculture*. Oxford and IBH Publishing Co. Pvt. Ltd., New Delhi, 193 pp.

Santhanakumar, G. and Selvaraj, A.M., 1995. *Concepts of Aquaculture*. Meenam Publications, Nagercoil, 204 pp.

Stickney, R., 1979. *Principles of Warmwater Aquaculture.* John Wiley & Sons, USA, 375 pp.

Yadav, B.N., 2006. *Fish and Fisheries* 2nd Revised and Enlarged edn. Daya Publishing House, New Delhi, 366 pp.

Yadav, B.N., 1995. *Fish Endocrinology*. Daya Publishing House, New Delhi, 170 pp.

Index